AF833466

Presented Discourse in Popular Science

Studies in Pragmatics

Series Editors

Maj-Britt Mosegaard Hansen (*University of Manchester*)
Kerstin Fischer (*University of Southern Denmark*)
Anne Barron (*Leuphana University Lüneburg*)

VOLUME 17

The titles published in this series are listed at *brill.com/sip*

Presented Discourse in Popular Science

Professional Voices in Books for Lay Audiences

By

Olga A. Pilkington

BRILL

LEIDEN | BOSTON

The Library of Congress Cataloging-in-Publication Data is available online at http://catalog.loc.gov

Typeface for the Latin, Greek, and Cyrillic scripts: "Brill". See and download: brill.com/brill-typeface.

ISSN 1750-368X
ISBN 978-90-04-36596-4 (hardback)
ISBN 978-90-04-36597-1 (e-book)

Copyright 2018 by Koninklijke Brill NV, Leiden, The Netherlands.
Koninklijke Brill NV incorporates the imprints Brill, Brill Hes & De Graaf, Brill Nijhoff, Brill Rodopi, Brill Sense and Hotei Publishing.
All rights reserved. No part of this publication may be reproduced, translated, stored in a retrieval system, or transmitted in any form or by any means, electronic, mechanical, photocopying, recording or otherwise, without prior written permission from the publisher.
Authorization to photocopy items for internal or personal use is granted by Koninklijke Brill NV provided that the appropriate fees are paid directly to The Copyright Clearance Center, 222 Rosewood Drive, Suite 910, Danvers, MA 01923, USA. Fees are subject to change.

This book is printed on acid-free paper and produced in a sustainable manner.

To Ace, Alexander, and Natalia

Contents

Preface

This book is an exploration of presented discourse in non-fiction. Using a corpus of popular science narratives of discovery, the study develops new approaches to functions presented voices and thoughts perform outside of literary fictions. As a linguistic exploration of popular science in book form, this work provides further support for the celebratory nature of popularizations and outlines specific mechanisms which create the positive picture of the scientific community.

One of the through lines of the study is the suggestion that popular science is strongly connected with fictionality, both the literary kind of fictionality and the kind of fictionality used in scientific work. Presented discourse of scientists plays a major role in fictionalizing popular science. Presented voices make it possible to dramatize scientific debates that took years to play out in print as dialogic exchanges. Presented discourse gives popular science authors an opportunity to showcase their human subjects as creative and caring communicators. Most of all, however, presented discourse contributes to establishing science as a domain of infallible geniuses.

Using current approaches to presented discourse analysis combined with structural investigations of narrative, this book offers a comprehensive look at what makes popular science books appealing to a large readership. *Presented Discourse in Popular Science* includes a theoretical and analytical portion (chapters 2–7), which examines the issues of fictionality, emotionality, and dramatization in fiction and non-fiction; supplies quantitative and qualitative assessment of presented discourse corpus data along with a comparison with other studies; and concludes with an examination of presented discourse outside the narratives of discovery. The last portion of the analysis confirms the phenomena observed in the narratives and reveals a unique feature of presented discourse outside the narratives—the fictionalized reader—a fictional actant created using discourse presentation.

In addition to theoretical and analytical chapters, this study includes an overview of the relationship between readers, writers of popular science, and members of the scientific community as well as a brief history of popularization (chapter 1) and practical suggestions for popular science writers (Appendix).

Detailed analysis of presented discourse, as offered here, often calls for examples that show multiple interactions of linguistic mechanisms. In order to guide the readers' attention to relevant portions of the sample texts, colored typeface is used throughout the book. Whenever only one type of presented discourse is exemplified, the highlighted text appears in blue, as in the following illustration of Narrator's Presentation of Thought Acts:

> On the same walk, Wilczek also came up with the idea for the "axion," a hypothetical low-mass cousin of the Higgs that is now a promising candidate for constituting the dark matter in the universe.
>
> CARROLL 2012: 169

Whenever the reader's attention needs to be directed to several presented discourse types, multiple colors are used. When this happens, the text immediately following or preceding the example supplies an annotation that explains the use of colored typeface.

Consider the following example of Indirect Speech (reporting clause is in green, Indirect Speech is in blue):

> Two thousand years later, the physiologist Leonard Hill argued in the 1920s that they [colds] were caused by walking outside in the morning, from warm to cold air.
>
> ZIMMER 2011: 10

Presented Discourse in Popular Science is designed primarily for linguists interested in presented discourse analysis; however, its opening chapter and the Appendix might appeal to a broader audience who wishes to see how linguistic analysis can be applied to the task of writing popular science and what investigation of language can reveal about the nature of popularizations.

Acknowledgements

In the writing of this book I am indebted to many people who supported and advised me.

In the United Kingdom, I wish to thank Professor Susan E. Hunston, OBE (University of Birmingham), who was instrumental in helping me develop and shape this research topic and who read many earlier versions of some of these chapters. Her commentary was invaluable and ultimately made this research project much better. To Professor Michael Toolan (University of Birmingham) I am grateful for the insightful suggestions and feedback on the material that appears here. Professor Greg Myers (Lancaster University) was kind enough to offer words of encouragement—for this I owe him great thanks.

In the United States, my thanks go to Dr. Ace G. Pilkington, my husband, who always takes an interest in and supports all of my academic endeavors. To him I am also grateful for feedback and proofreading; I, of course, take full responsibility for any remaining errors. To the library staff of Dixie State University and Brigham Young University I extend my thanks for supplying many of the sources cited in this book.

In Russia, I wish to thank Dr. Alexander Ivanchenko and Dr. Natalia Ivanchenko, my parents, for their advice on the inner workings of the international scientific community.

In the Netherlands, I am thankful for the support of the Brill publishing team I worked with: Irene van Rossum, Maarten Frieswijk, and Fem Eggers. Their insight and suggestions were immensely helpful for navigating the publishing process. I also extend my thanks to the anonymous scholar who conducted the peer review and whose comments were instrumental in the final organization of the book. To the typesetter TAT Zetwerk who worked on this project I am grateful for the extra attention to the details.

Last but not least, I extend my gratitude to the scholars and scientists whose prior research made this project possible.

List of Tables

List of Abbreviations

DS	Direct Speech
DT	Direct Thought
DW	Direct Writing
(F)DS	(Free)Direct Speech
(F)DT	(Free)Direct Thought
(F)DW	Free Direct Writing
FIS	Free Direct Speech
FIT	Free Indirect Thought
FIW	Free Indirect Writing
IS	Indirect Speech
IT	Indirect Thought
IW	Indirect Writing
N	Narration
NI	Internal Narration
NPSA	Narrator's Presentation of Speech Acts
NPTA	Narrator's Presentation of Thought Acts
NPV	Narrator's Presentation of Voice
NPW	Narrator's Presentation of Writing
NRS	Narrator's Report of Speech
NRT	Narrator's Report of Thought
NRSA	Narrator's Representation of Speech Acts
NRTA	Narrator's Representation of Thought Acts
NRW	Narrator's Report of Writing
NRWA	Narrator's Representation of Writing Acts
NV	Narrator's Representation of Voice
NW	Narrator's Representation of Writing

CHAPTER 1

Introduction: Popular Science, Overview of the Genre

Why Popular Science is Important and How it Shapes the Reader's Perception of the Scientific Community

If asked why popular science is important, an average reader might identify its ability to relate the latest scientific knowledge in a way that does not require a PhD in physics or astronomy, for example. She might add that in a science- and technology-centered society it is necessary for the public to have some basic idea of what makes the world go round—literally and figuratively. A reader could also point out that popular science is the only way he can connect with the scientific community and that interactive popularizations such as blogs or on-line articles that allow commentary provide an outlet for him to voice his opinion and evaluate the latest breakthroughs on an emotional, ethical, or personal level. Yet another group of readers might argue that popular science has attained a religion-like quality, with publications that promise the answers to the mysteries of creation, parallel worlds, time travel, eternal life, and similar technomiracles.

All of these answers will be correct, and numerous studies in linguistics, sociology of science, and history and the philosophy of science will verify the hypothetical responses I introduced above. For a brief excursion to the scholarly evaluations of popular science's importance, consider a short list of fairly recent studies: Turney 2007, Luzón 2013, Lightman 2000, Laslo et al. 2011, Bensaude-Vincent 2001, Supper 2014.

However, neither the scholars nor my hypothetical readers reflecting their findings suggested what I think is very important and just as obvious about popular science as the commonly supplied answers: popular science is important because it is an outlet of the scientific community that could be more readily and comprehensively evaluated by an outsider than any other manifestation of research or written artifact that comes out of the laboratory.

So far, the scholarly evaluations of popular science by linguists, sociologists, historians, and philosophers have focused on the differences between professional scientific writing and popular scientific writing (see, for example, Turney 2004b, Topham 2000, Stanley 2011, Luzón 2013, Livnat 2012, Lightman 2000, Laslo et al. 2011, Hyland 2010, De Oliveira and Pagano 2006, Ciapuscio 2003,

© KONINKLIJKE BRILL NV, LEIDEN, 2018 | DOI: 10.1163/9789004365971_002

Calsamiglia 2003, Bensaude-Vincent 2001). Of course, the language of a popularization is different from the professional language in many ways; the studies mentioned above would confirm that. Many of them will also tell you that the difference in language is determined by the difference in audience or setting. For instance, De Oliveira and Pagano (2006: 628) contrasting popular and professional scientific articles talk about "different settings in which knowledge circulates."

The attention to language is easy to explain. As Hyland (2010: 117) suggests, "language becomes a form of technology, or a resource for controlling an environment, by presenting interpretations and positioning participants in particular ways to establish knowledge." In other words, language is a powerful tool for scientists who can use it to control who has what degree of access to scientific knowledge. Write a research article, and you have disseminated your findings among your fellow scientists and gained prestige for being able to code your results in such a way that only a select few can decipher. Produce a popularization, and you have enlightened the lay masses. Both writings would be concerned with the same scientific knowledge, but each one would frame it differently.

In our days, the idea that there is a strict divide between the scientific community and the lay public has been challenged. And several linguists have demonstrated that the very differences in the language of professional and popular science led to a realization that "'science' is not a monolithic entity always understood in the same way, but a social construct created by different groups with different interests" (Hyland 2010: 118). Myers (2003: 267) puts the same idea this way: science is not "a discourse, a single set of social practices around one thing, but … a terrain of competing discourses and practices." As such, both Hyland (2010) and Myers (2003) recognize that popularizations themselves are written for various audiences. Books "by scientists for an elite educated audience," and articles in magazines, newspapers, and blogs for a less sophisticated "public" (Hyland 2010: 118). Yet, there is one more audience for popular science, and that is the members of the scientific community themselves.

Consider this, it takes months of reviews and revisions to publish an academic article and only a matter of hours to post your findings on a blog or to call a press conference. You have probably heard of the 1989 press conference at the University of Utah during which Pons and Fleischmann announced that they had mastered cold fusion. (They were wrong.) "Scientific discourse at the public level is only apparently 'public': communication at this level is not actually meant to address the general public, but to reach a vast number of colleagues rapidly … without having to conform to the times and constrains of specialist communication" (Bucchi 1998: 12). There is also always the possibility of getting "scooped" by a rival team of researchers while waiting for the article to be peer-reviewed, printed, or posted online, not to mention the politics involved with the acceptance to the major or "luxury journals" as Randy Schekman calls them.

Schekman (2013), a winner of the Nobel Prize in Physiology or Medicine in 2013, has been speaking out publicly (that is through the popular science sections of newspapers, magazines, and in online publications) against journals that set out to "make waves because they explore sexy subjects or make challenging claims" rather than present "important work, such as replication studies." He encourages his colleagues who read the online edition of *The Guardian* to disseminate their research findings not through *Cell* and *Nature* (the prominent journals in the fields of biology and genetics, among others) but "through the new breed of open-access journals that are free for anybody to read, and have no expensive subscriptions to promote." However, as Bucchi (1998) and Turney (2007) show, popular science texts have also become alternative outlets for some researchers.

The examples abound. One of the less known of them took place in 1967 when Bryce DeWitt, a renowned physicist, decided to revisit the largely forgotten idea of a multiworld first proposed by Everett in his 1957 PhD thesis. As Brian Greene (2011: 190–191) tells the story, "DeWitt wrote a general level summary for *Physics Today* [a popular science magazine—O.P.] that reached a much broader scientific audience." From that point on, the theory was no longer marginalized. Not only does popular science address professional scientists, but it also serves as a medium for negotiation and development of ideas, as much as the research article does. As Turney (2007: 2) notes, "Some of the physicists [sic.] arguments—shorn of the maths—are being fought out in popular books. Proponents of superstring theory such as Brian Greene are under attack by those who favour other approaches like Lee Smolin partly because they have been so good at promoting their ideas to the wider public. Smolin's recent *The Trouble with Physics* is an attack on PR as much as on superstring theory." Bucchi (1998: 11) cited a study that "shows that articles appearing in *The New England Journal of Medicine* are cited twice as much by specialists if they are also mentioned in a daily paper like the *New York Times*."

As Hyland (2010: 118) notes, "It would be a great oversimplification to dismiss popular science as merely infotainment. This is a discourse related to the academy, its work, and its forms of communication." One of the side effects of the professionalization of science is increased specialization. A physicist and a marine biologist are both scientists, but one will not be able to grasp all the intricacies of the other's research without the help of popularization. Moreover, even a professional in one field might not be familiar with the advancements and recent developments taking place outside of his immediate area of expertise.

Myers (2003: 268) provides an excellent example, "When I go to the doctor, I treat her as an expert in medicine, but her relation to current medical research will generally be as a continuing student, not as a participant, and the medical journals have to perform a kind of popularizing function for her." However, she might not be entirely satisfied with the professional publications or not have the time to sort through them to find the information that will have the most impact on the field. In this case, popular publications will act as a filter for her and offer the most striking recent findings and breakthroughs.

In these circumstances, it is imperative to look at popular scientific writing with the same level of scrutiny as the scientific community affords its professional publications. Popular science is important because it opens up the scientific community to a level of examination otherwise not accessible. As Carl Djerassi (1994: xiii), a chemist and the inventor of the birth control pill writes, "Science is conducted within a close-knit culture whose members are generally reluctant to disclose their tribal secrets." However, popular science gives us an unprecedented exposure to the world of science and reveals how its members develop, produce, negotiate, and ultimately present what we call the scientific truth.

The authors of both popular and professional scientific texts engage in writing that in addition to delivering facts is designed to shape perceptions and alter reality. Even though Turney (2007: 2) calls popular science "a literature of reality, of *how things are*," the more accurate description of the genre would be a literature of how the scientific community wants the public to see the scientific reality, of how things should be. Turney (2007: 2) later suggests that "we should examine how a [popular science] book treats the nature of science." I propose that we should also ask what is the nature of science? The answer lies, among other places, in the approach to language undertaken by practitioners of science through the centuries.

History of Popular Science

Language: Professional and Popular

In order to understand the language resources of popular science and their application, it is imperative to consider them in conjunction with the language of professional science. Looking at the language of popular science in isolation from the language of the professional scientific community strips away a vital portion of the context in which popular science exists. Popular scientific writing should not always be evaluated as an entirely separate discourse category but considered part of the continuum that is scientific writing. As Myers (2003: 267) puts it, science, from a linguistic point of view, is not "a single set of social practices around one thing, but … a terrain of competing discourses and practices." Popular science undeniably represents one of these discourses and activities.

The scientific community has a history of manipulating the language that is used to communicate their findings not only to the outside world but to each other. They do this in order to preserve the privileged social position of science and, to some extent, to create a linguistic identity that distinguishes scientists from other members of society and prevents easy entrance into the scientific community. One way to maintain such an exclusive intellectual community is through the control of the language that it uses. For a while now, English has been accepted as the international language of professional science. As Gordin (2015: 2) puts it, scientists are "the most resolutely monoglot international community the world has ever seen." He adds, "the exclusive language they use to communicate today to their international peers is English" (Gordin 2015: 2).

Historically, however, English has not always been the language of science even in the English-speaking world, let alone internationally. Latin performed that function for European scientists from the middle ages up until the mid-seventeenth century. This is a well known story, but what's left out of it quite often is the role that English played as a language of popularization (in the English-speaking world) even at the time when Latin was the lingua franca of professional science. Circa 1391, Geoffrey Chaucer wrote *A Treatise on the Astrolabe*. In the words of Seth Lerer (2015: 80) it was "a synthesis of medieval astronomical and astrological teaching inherited from Greek and Latin, Arabic and European teachings" with the idea of simplifying "a technical language." In essence it was a popularization. The audience for the *Treatise* was Chaucer's ten-year-old son, Lewis, to whom Chaucer hoped to "reveal … in … easy English the conclusions concerning this material in as true a fashion as any ordinary treatise shows in Latin" (transl. from Middle English by Lerer 2015: 81). In the introduction to his text, Chaucer makes it clear that he is using English "only for [his son's] instruction" (Lerer 2015: 82). In doing so, Chaucer recognizes a basic goal of a popularization and the inaccessibility of the language of science to those who are not specially trained in its usage—very modern issues.

Later on, when the English language established itself as the language of all the aspects of English society, including science, scientists set out to find a way to induce the separation once again. The fact that we are discussing the "language of science" as a category suggests that such a language can be separate from either everyday discourse or from a native language/s one speaks. Latin as a language of science is an example of the latter, and the discourse of popularizations analyzed in this book illustrates the former. As Gordin (2015: 24) writes, "'Scientific languages' are either specific forms of a given language that are used in conducting science, or they are the set of distinct languages in which science is done." In any case, the idea of a language used for science as different is by now universally recognized. But who is responsible for establishing and maintaining the differences in the English language?

From the onset of professionalization in the eighteenth and the nineteenth centuries to the present day, there is evidence of scientists themselves shaping the image of and controlling the access to scientific findings. Professional science as a social category was virtually non-existent until the nineteenth century. Before then, the sciences were accessible to lay persons and, in fact, attracted many amateurs who ended up making scientific contributions. The end of the nineteenth century in Europe, Britain, America, and Russia saw a remarkable vitality of the print market, a large portion of which included scientifically-themed texts (Banerjee 2012: 8, Pilkington, A. 2017: 13, Topham 2000: 561). These publications, while disseminating scientific information, were designed to create armchair scientists, who did not venture outside the comforts of their reading rooms. And while the process was gradual, as Topham (2000: 561) notes, this was one of the contributing factors to the creation of a passive lay public as opposed to a group of enthusiastic amateurs and professionals who essentially worked side by side.

The alienation between the public and scientists can be attributed to many causes (the increasing complexity of research procedures, the desire of the professionals to monopolize scientific knowledge, etc.) and even to a specific person—T.H. Huxley.

As Bernard Lightman explains, "Huxley and his allies ... worked to purge scientific societies of wealthy, aristocratic amateurs, Anglican clergymen enthralled by natural theology and women with a keen interest in science" (101). The inclusion of amateurs was, according to Coppola (2016: 3), a calculated economic and political move on the part of Charles II, who "saw the potential for real material benefits, or at the very least an inoffensive diversion, in the tireless and omnivorously curious pursuits of the so-called virtuosi, the non-professional gentlemen-scientists who filled the ranks of the Royal Society and read the *Philosophical Transactions*, amateurs who experimented for no reason other than a virtuous love of truth, and to beguile their leisure hours in ingenious pursuits." Huxley's goal was to establish professional scientists as the only keepers and distributors of Natural knowledge.

These actions became known as boundary work—a term that is still used today to describe focused efforts to discourage public participation in and influence on science. As Lightman (2000: 101) sees it, scientists of the 1890s tried to secularize science by separating it from what he calls "elements that previously had connected public and scientific culture, including anthropomorphic, anthropocentric, teleological and ethical views of nature." The professional scientists of the second half of the nineteenth century set out to establish themselves as the only authorities on secular, natural knowledge.

Another explanation for the separation of professional science from the public is connected with the dramatic change in the complexity of scientific procedures and the new apparatus involved that now required specific expertise more so than general experience and curiosity. Topham (2000: 560) calls this process a series of "epistemological and rhetorical shifts" that positioned scientific discoveries "as the preserve of scientific 'genius'" rather than as an activity open to anyone.

While the second explanation does not blame any one scientist in particular, it still holds the scientific community responsible for the boundary work and the general outcome as a negative occurrence. There is, however, a different view of the situation. Bensaude-Vincent (2001) suggests that the separation between the scientists and the public is not as unnatural or as detrimental as Lightman or Topham see it. For her, the divide between the professional scientists and the lay public is "a necessary ingredient of our [western] notion of science" (Bensaude-Vincent 2001: 100). The animosity between the two groups, on the other hand, she does regard as unnecessary and harmful. For Bensaude-Vincent (2001: 105), the gap is not a static notion of separation but a process of alienation which inevitably involves the manipulation of language.

Whatever the reason, once the access to scientific laboratories was denied to all but a select few, Huxley and his like-minded colleagues set out to create and promote a specialized, jargon-laden language designed to code the scientific findings sufficiently so that only the members of the inner circle would be able to make sense of them. Topham (2000: 561) calls the emergent language of professional science "increasingly arcane and technical." Lightman (2000: 101) describes the language of the scientific professionals of the time as "detached …, heavily spiced with complex scientific terms." In addition to the exclusion of the non-professionals, such use of language deliberately expunged story telling from professional scientific writing, making scientific results appear as entirely objective and untainted by human interference. (The use of narrative structure, however, has never been entirely abandoned by professional scientists and has since gained recognition in popularizations. See Appendix.) With a linguistic barrier in place, it was now possible to present science to the masses in a controlled way that did not threaten the hegemony of the professionals.

According to the *Oxford English Dictionary*, the verb "to popularize" acquires the meanings that pertain to the diffusion of scientific knowledge at the very time when science was emerging as an exclusive domain of the professionals. The first mention of the word "scientist" is in 1834. The idea of making complex information accessible, interesting, and appealing predates the name for a professional natural philosopher by about thirty-five years, but it probably has never been as essential as it became at the time when the language of professional science was being created.

Having succeeded in eliminating the involvement of amateurs and creating a specialized language of science, Huxley and his followers (John Tyndall, for example) attempted to control how scientific findings were delivered to the lay masses. The new popularity of the print medium (especially books) provided a welcome dissemination channel. However, as Topham (2000: 600) rightly points out, once the scientists ventured to popularize through print texts as opposed to public lectures, for example, they no longer had the same level of control in the message that they presented. It was a lot easier to engage in creating the positive image of science through popular talks than it was to deliver the same optimistic message in print. As Coppola (2016: 9) demonstrates, the connection between science and visual presentation of new knowledge and new technologies was a legacy of the eighteenth century popularizing mechanisms. "Science was turned to spectacle," he notes, and "unambiguously offered as entertainment," which was conceived, produced, and performed by scientists themselves. Henry Pepper's ghost lecture at Royal Polytechnic Institute is a representative example: "Henry Pepper conjured up on the stage by virtue of an elaborate hidden optical apparatus" a projection, which he "would exhibit … with pomp and mystery, in a lecture that subsequently divulged the secrets of the contrivance to the audience. Then … Pepper would conclude the evening's entertainment by featuring his ghost as a show-stopping special effect in the performance of a gothic play or popular opera" (Coppola 2016: 8). Such spectacular means of popularization were not available through print media and the whole concept needed to be overcome in order to preserve science as an elevated activity of a select few.

The new, textual channels of popularization brought in new challenges. One of them was the loss of complete control of the shape of the message and dissemination mechanisms. "Publishers were not merely the lackeys of scientific practitioners. They were also extremely important initiators of publishing ventures in which scientific practitioners featured mostly only secondarily, and often not at all. As the creative agents behind vast numbers of popular trade books and periodicals, they were primarily responsible for a large proportion of the printed matter on natural subjects" (Topham 2000: 600).

As Topham (2000) acknowledges, there is a relative lack of knowledge about the activities of the publishers of popular science at the time of professionalization. His research, however, shows that commercial publishers, who sought to make money and therefore depended on the volume of sales, as opposed to subscription publishers, who took fewer risks, dominated the print market. "For, while private and subscription publication continued to be possibilities for authors whose manuscripts were turned down by commercial publishers, the growing complexity of the book trade in the early nineteenth century made it progressively more necessary that a book should have the backing of a commercially motivated publisher if it were to succeed" (Topham 2000: 581–582). The publishing trade then, just as the publishing trade is now, was motivated to make maximum profit. "The competitive commercial climate arising both from the growing demand for print and from the removal of perpetual copyright … served to encourage booksellers to be innovative in seeking to exploit new markets" (Topham 2000: 580). And popular science as a genre that was emerging in its new form at that time provided access to one of those new markets.

Readers always enjoyed good stories, and science presented in narrative form is always appealing. As Olson (2015: 9) puts it, "The journals want to tell good stories, the scientists want to tell good stories, … and the journalists want to tell good stories." However, in eighteenth and nineteenth century Britain, the time when professional scientists were inventing their own professional language and consequently the language of popularization, narrative was not the preferred form of organizing scientific information for the public. It was dialogue.

Popular science at that time was offered to the lay public in a form that confined thought experiments, analogies, and actual scientific demonstrations to the familiar surroundings of an average Victorian home. It was familiar science (Keene 2014). Popular science became the science of the home, and was very different from the "real" science that was to be conducted by professional scientists in their laboratories. Such division of the domains proved to be one more means to secure the separation of professional and popular science—one is done in a laboratory with the help of specialized apparatus, while the other can take place in a kitchen and uses everyday household objects. This phenomenon produced a kind of paradox where the scientists were eager to explain their actions to the public, but in doing so distanced their audiences from the "real" work. It was common to regard such an approach to popularizing scientific research as a translation—a method of popularization that later became known as the dominant model. It presupposed the nearly-total ignorance of the public in all matters scientific and certainly did not welcome or supply any opportunities for constructive feedback from the lay people.

Popularization materials of the time were commonly organized as dialogues; it proved to be a successful publishing practice (Myers 1992, Keene 2014: 55). Jane Marcet's *Conversations* on the subjects of chemistry, natural philosophy, and vegetable physiology are good examples from the period. Dialogue was seen as an effective educational strategy, presenting complex scientific knowledge in a non-intimidating form. As Keene (2014: 55) puts it, dialogue "represented actual conversational possibilities" and in print form was a logical extension of the lectures and talks given to the public. Myers (1992: 238) sees dialogic form as "an ideal of discourse" that offers to the reader a perfect set up for a discussion of any scientific question. Printed dialogues as a means of popularizing science, as Myers (1992: 238) points out, bring to the forefront the issue of the scientific language by focusing on explanations of terminology and making sure the reader understands the basic terms before moving on. By doing that, dialogue acknowledges the difficulties the language of professional science poses for lay readers. Other popular texts do that as well, but, in the words of Myers (1992: 238), dialogues "teach us to read conventional scientific texts by making explicit what is usually implicit … stepping back from the text involved in dialogue allows for discussion of terms … take for example … Mrs. Marcet's … discussions of terms."

These "discussions of terms" were increasingly necessary as the scientific community was deliberately setting itself up as a separate linguistic, as well as intellectual, entity. Those popularizers who were outside the elite circle of professional scientists had to learn and practice the new language along with their readers. As Lightman (2000) explains, this was not altogether a bad thing for those interested in science. Unable to practice but still possessing basic scientific knowledge, the groups excluded from the professional scientific community (women, clergymen, wealthy aristocrats) found new applications for their scientific passions. As Lightman (2000: 101) puts it, "If it was the age of professionalization, however, it was also the age of the popularization of science."

Jane Marcet is an excellent representative of the interested public excluded from the professional scientific community. Myers (1997: 45) makes an important point about Mrs. Marcet: "Unlike her contemporary (and friend) Mary Somerville, she did not see herself as contributing to the fields about which she wrote; she saw herself as a popularizer of the work of others." While Myers (1997: 45) is quick to note the modesty of her view, it is telling that she accepted her role as a popularizer, not a scientist. Marcet, however, was not completely without access to the professional scientific community, as her husband was a doctor "known for his analysis of spa water and who later lectured in chemistry" (Myers 1997: 45).

Lightman (2000) attributes the high numbers of popular scientific texts being produced at the end of the nineteenth century to the advances in printing technologies and to the forced emergence of trained scientists who were no longer able to contribute their knowledge and skills in any meaningful way during the age of the new, professionalized science. At the same time, the role of the language of professional science should not be overlooked. Keene (2014) and Myers (1992) show that the end of the nineteenth century was the age when popular scientific dialogues prevailed, and both stress the simplicity with which popularizations structured as dialogues teach the new way of talking about and practicing science.

The popularity of dialogic form decreased as time went by, however, as Myers (1992: 222) observes with some surprise, "Many authors have turned to dialogue or something like it in the twentieth century." He attributes the popularity of this form of knowledge presentation to dialogue's ability to bring to the surface the argumentative and communicative nature of science. On the other hand, Myers (1992: 238) notes an important feature of dialogic presentation that he regards as negative but which may be a strong contributing factor to dialogues' longevity as a form of popular scientific discourse. Myers (1992: 238) writes, "There is a way in which dialogues stifle rather than promote the dialogical. They mark out and complete the imaginable space of discussion. They lay out the possible positions and then take control of them all." What Myers means is that a print dialogue presents, perhaps, the most author-controlled mechanism of delivering information, more so than narrative. A narrative is mainly the voice of the author, while a dialogue requires also the voice of the reader. The reader in such a text becomes a construct which is restricted to a series of questions and answers predetermined by the author. The rise and the proliferation of popular science in dialogic form in the eighteenth and nineteenth centuries may be explained by the language-controlling properties of that medium. Not only does it give the author a chance to tailor the language she/he is using to address the public, but it also affords an opportunity to demonstrate the kind of scientific language (or lack thereof) that is acceptable for use by the lay public.

Those who write scientific texts at any level (popular or professional) have been willing and active manipulators of language. It is clear that professional scientists are not the only ones responsible for the linguistic legacy of science. Publishers and popularizers have played their roles along the way, distributing, promoting, and ultimately making acceptable the kind of language offered by the scientific community.

Authors and Readers

Modern popular science texts do not use dialogue as the preferred form, but they do emphasize dialogic exchanges through the use of presented discourse and also maintain a kind of reader control common in dialogues. In chapter 6, I discuss the fictionalized reader and the types of presented discourse used to create such a textual entity. The fictionalized reader found in contemporary popular science is very similar to a type of fictional reader introduced through dialogue. However, the reader present in popular science dialogues was always a fully-fledged fictional character, while the fictionalized reader of today is only a semi-fictional representative of the audience. In both cases, however, presented discourse plays a crucial role in characterization of this textual participant.

The fictionalized reader is a less author-controlled option than the entirely fictional character employed in dialogues, but it is still, perhaps, a remnant of the past attempts to determine not only the content and the language of popularizations but also to shape the reading public as a group that stands in opposition to professional scientists.

Once the access to professional science was denied to the public, popularization of science became more in demand than ever. The public's interest in all scientific matters remained strong throughout the years of professionalization of the scientific community. In terms of linguistic analysis, the popularization model that emerged as a result of the gap between the professionals and the lay people suggests the presence of three distinct discourse types: the discourse of professional science, the discourse of the lay public, and the mediating discourse of popularization. The discourse of popularization was often approached as a kind of translation—this, perhaps, was an especially apt metaphor at the time of professionalization of science, when a new scientific language was asserting itself. However, the view of popular scientific discourse as a "translation," or worse a "distortion" has persisted into the twentieth century (Lightman 2000: 100; Moirand 2003: 175, 176). Moirand (2003: 175) called such popularization structure a "triangular arrangement" or a "triangular communication model," referring to the three discourses involved. She also observed that in these circumstances, "the popularization becomes the work of … [the] third actor"—the popular science author.

Moirand's (2003) assessment falls in line with Lighman's (2000) observations that the professional scientific community disengaged from the public and relegated the popularizing tasks to those who did not belong among the scientific elites. Such disengagement from popularization inevitably led to a loss of connection with the public (which modern scientists are desperately trying to repair). Popular science and its consumers were seen as passive and worthless. However, recent research into the popular scientific culture and its products (e.g. popular science texts) suggests that the arrangement was never that simple nor has it been simplified over time. Looking back, historians of science are starting to notice that popular science "can actively produce its own indigenous science, transform the products of elite culture in the process of appropriating them and substantially affect the nature of elite science as the price of consuming the knowledge it is offered" (Lightman 2000: 100). Examining the modern state of popular science affairs, researchers are noting the increasingly influential role not only of the popularizer but of the reader as well.

The public, it seemed, kept resisting the image of the scientifically ignorant masses assigned to it. It is only logical that sooner or later a different view of popular science, the one that recognized and acknowledged public input, would emerge. The perception of the role of popularizations as translations was eventually overcome. The works of Myers (1990, 2003) and Bucchi (1998) played an important role in changing how popular science was regarded and how it should be written.

Moirand (2003) demonstrates that the content of contemporary popular science, to some degree, is determined by the reader—and this does not have to do with the popularity of certain subjects over others. Popular science, as Moirand (2003) and others (see, for example, Turney 2004b, Herman 2009) assert is primarily concerned with providing and explaining scientific information. At the same time, popular science tends to cover material that is new or research that is still under investigation; in these cases, science does not yet hold the answer, and that makes the traditional goal of explanation difficult to fulfill. As Moirand (2003: 187) observes, "It is not easy to 'explain', describe or name, even, relatively unknown or contested scientific or technological phenomena. As a result, these discourse moments, caught between science and politics, the technological and economic worlds, come to encourage the emergence of new objects of discourse, which are not only 'scientific' or 'technical' in nature." The new concepts talked about in popular science are of a social nature as much as they are scientific or technical.

The readers, it turns out, have over the years created a demand for an alternative to scientific explanations when those are not available. As Moirand (2003: 191) shows, "scientific explanation makes way, then, for explanation in terms of the social meaning of the events in question." In other words, popular science is expected to supply not only facts but also evaluations of the information in terms of the social consequences of research.

Moirand (2003) suggests that this focus on the social meaning of science creates an additional layer in the traditional model of science communication,

"A new type of discourse on science has appeared, which, rather than simply replacing the existing one …, exists alongside it, doubling it up" (176). Myers' (2003) insight supplies a plausible explanation for how this additional type of discourse (explanations of social consequences) came to exist. While the general public might be ignorant of "any specific piece of scientific information, however basic," people tend to know the science behind the issues that affect their daily lives. In Myers' (2003: 268) words, non-specialists know "about the areas of science and technology that concern them." He continues with the following examples, "The parents of a child with a rare medical syndrome may have a considerable knowledge of that syndrome. Opponents of nuclear power … have a considerable knowledge of the science of nuclear hazards and the technology of nuclear waste storage … Bantam breeders may know a great deal of practical genetics." These interests and this expertise translate into demands for specific coverage of science. Myers (2003: 269) points out that this kind of knowledge creates a powerful combination between "the abstractions of scientific knowledge" and "lived experiences." Armed with the evidence of social consequences, "members of the public have their own persuasive resources" (Myers 2003: 269). Thus a reader of popular science is no longer a passive observer but an active participant in the social discourse related to science and its consequences. That is why it has become extremely important for science to offer itself to the public in a positive light.

A Genre with an Agenda

Science as presented through writing, be it professional research articles or popular accounts, is often shown in an exclusively positive light and as an activity that never results in failures. Popular science adds to this list a focus on intellectual activities at the expense of descriptions of experimental procedures (Pilkington 2017a).

The fact that popular science presents scientists and their work in an exclusively positive light has been discussed before. Several studies make similar claims covertly or overtly: For instance, Fu and Hyland (2014: 123) argue that popular science is "persuasive, seeking to convince the reader both of the importance of the content and a wider ideology of scientific progress." Bucchi (1998) makes a less veiled claim that popular science employs what he labels "'celebratory' discourse"—essentially a celebration of science and its achievements and practitioners. Harré (1994) points out that professional science and the publications it produces are not much different—focusing on positive outcomes of laboratory procedures at the expense of the truth. Harré (1994: 87) writes, "If anyone tried to publish a story more like real life, in which hypotheses were dropped for lack of support, apparatus couldn't be made to work within the parameters of the original experiment, and so on, it would be turned down." He continues, "Science must present a smiling face both to itself and to the world" (Harré 1994: 87). And so it does. In fact, it has become so good at it that its professional and popular outlets have developed specific linguistic formulae for constructing "a smiling face." The use of narratives is one of them; presented discourse of scientists is another. In the Appendix, I explain how both of these mechanisms can be used by popular science authors to create texts that conform to the current parameters of how science should be popularized.

Including the voices of professional scientists into popularizations adds credibility and authority to the texts. However, a deeper examination of these voices and their functions offered in this book will demonstrate that they serve to enliven accounts of scientific discoveries through dramatization, enhance explanations of scientific concepts, and overall to further the celebratory exposition of science and its practitioners.

For instance, a majority of the voices introduced into the popular science books examined for this study dramatize the events they describe. This is true for all direct and indirect forms of discourse presentation. In fact, dramatization is so pronounced that even forms of indirect discourse typically used for their information-carrying properties exclusively contain a degree of dramatization in popular science. Dramatization through presented voices serves as an endearing mechanism that shows scientists as not only sensible and concerned with facts but also as emotional, creative, and relatable.

Presented thoughts of scientists, on the other hand, surprisingly, show them as thinking about science all the time. Nothing private or personal is ever revealed through relaying the thought processes. This is highly unusual since the prototypical function of presented thought is to introduce an intimate, inner world of a person. In popular science, however, this is the preferred method for describing scientific hypotheses and for announcing discoveries. Such treatment not only projects unrealistic levels of dedication, but also points to the scientists in the books serving a purpose beyond the introduction of scientific facts. Analysis of presented discourse shows the members of the scientific community to be carefully crafted characters akin to those found in fictional stories.

Olson (2015: 9) attributes the focus on positive presentation of science to the basic need of attracting audiences through good storytelling, which he equates with positive outcomes of scientific research, "The positive result is the same as telling a good story." Communication of science, Olson (2015: 10) demonstrates, favors the impact a story will have over the "soundness" of the research that went into a particular scientific result. That is certainly true, and in my own research I have pointed out popular science's preference for the sensational over the mundane (see, for example, Pilkington 2017a). Baron (2010: 201) mentions the propensity the media has to distort scientific facts in order to present sensational stories, "Scientific debates aired in the media are often focused more on the conflict and less on the substance." Olson (2015: 9) gives the following statistics from the area of health sciences in the UK: "In 2014 Petroc Sumner and his colleagues ... examined biomedical press releases from 20 major UK universities versus the published research papers upon which the releases were based ... 40 percent of the press releases contained exaggerated advice, 33 percent contained exaggerated causal claims, and 36 percent contained exaggerated inference." Molek-Kozakowska (2016: 65) reports similar findings and suggests that the celebratory discourse of much of popular science comes from "uncritically follow[ing] science PR and press releases."

Most often the issue of positive bias or the sensational view of science is labeled as a problem. Trying to present scientific results in an overly-optimistic light is misleading and certainly problematic for lay people who are trying to make sense of the research. At the same time, designating the positive bias in science as a problem is, on some level, akin to dismissing it as something not worthy of serious attention and investigation. From that point of view, the positive bias is something to get rid of, not something to analyze and study.

The emphasis on positive presentation of science is a phenomenon that does have some definite drawbacks, but it provides valuable and meaningful linguistic insight into the scientific community. It has become an accepted practice with a concrete set of linguistic features which need to be investigated since the majority of professional and popular texts rely on them heavily. These are deliberate and well-developed linguistic resources that in popular science represent definite patterns on the lexical and structural levels.

Linguistic research shows multiple locations for positive bias within popular science texts. For instance, Koteyko et al. (2008) and Moirand (2003) analyze vocabulary choices designed to promote the positive image of science. Fu and Hyland (2014) examine the role of reader-engagement techniques in presenting an optimistic image of science. Bucchi (1998) and Turney (2004a) show how certain metaphors contribute to the favorable perception of scientific activities by the public. It appears that a lot of effort has been devoted to producing a specific linguistic set of features that facilitate the emphasis on the positive outcomes of scientific research. These linguistic features are not random but a result of deliberate actions. That is why labeling the positive bias in science presentation as a problem is counterproductive as such labeling obscures the real issue—the focus on positive outcomes and the celebration of scientific endeavors is a major goal of popular science and, to some extent, of professional science as well. Whether we like it or not, this is what popularizations set out to do.

Conclusion

As a genre, modern popular science emerged as a response to social changes. Today, it continues to be at the forefront as our society adjusts to the increased influence of technology. As I mentioned at the beginning of the chapter, the popularity of the genre has resulted in an almost religion-like following among some of its readers.

For many, popular science has become synonymous with science in general, and the crossover between scientists and authors makes it easy to make the connection. However, it is a mistake to assume that the influence of popular science ends with the presentation of scientific facts. If the history of the genre and its present state teach us anything, it is that popular science is as much a social phenomenon as it is a scientific outlet. To allow for professional science to appear agenda free, popular science assumes the role of promoting, supporting, and recruiting for the professional scientific community.

These goals are achieved in a variety of ways. One of them is the time-proven approach of language manipulation. In contemporary popular science it manifests as specialized use of presented discourse and fictionality which result in a hybrid genre incorporating non-fiction and fiction-like elements. The impersonal contributors to the research article become engaging characters on the pages of popular science books. Their research findings—often outcomes of monotonous laboratory experiments—spring to life as exciting stories of discoveries that come about through sparks of genius. Reading popular science is finding out about important scientific developments, and it is also accepting the infallible nature of science itself.

CHAPTER 2

Theoretical Perspectives

Presented Discourse: An Overview of Analytical Approaches

I will begin the introduction to the basics of presented discourse analysis with Leech and Short's (1981) influential model that has become the standard for analyses of presented discourse and serves as the foundation for the subsequent models I will address. Next, I turn to a discussion of Semino and Short's (2004) linguistic study that examined presented discourse in fiction and non-fiction. Their study is the most comprehensive one at the time of writing and has introduced new terminology and analytical categories that expand the Leech and Short model. I will give more details of the Semino and Short approach in Chapter 3, as I compare my own findings with their data. I will also refer to their conclusions throughout this book to show how they relate with my own results. The overview of the Semino and Short framework is followed with a summary of Short's two more recent publications on the subject of presented discourse analysis—Short (2007) and Short (2012)—in which he offers additional insight into the earlier models as well as proposes some changes in terminology.

The models I discuss in this section contain all the analytical categories used for presented discourse analysis. Any other approach is a modification of one or the other of these frameworks, as Short's later works demonstrate. The following subsections are designed to provide a concise, yet comprehensive summary of discourse presentation analytical models to date. As such, the frameworks introduced below demonstrate slightly different analytical angles while holding true to the common core. The information below is intended as an overview or an introduction to presented discourse analysis. While I use the examples from the texts I analyzed as illustrations here, my own model of presented discourse analysis will be given in Chapter 3. Before I introduce each model in detail, I would like you to consider terminological and functional comparisons of the analytical frameworks. See table 2.1 and table 2.2.

© KONINKLIJKE BRILL NV, LEIDEN, 2018 | DOI: 10.1163/9789004365971_003

TABLE 2.1 *Terminological comparison of presented discourse analytical models*

	Designation of model			
Type of presented discourse	**Leech and Short (1981)**	**Semino and Short (2004)**	**Short (2007)**	**Short (2012)**
Presentation of Speech	Direct Speech, Free Direct Speech, Indirect Speech, Free Indirect Speech, Narrator's Representation of Speech Acts	Narrator's Report of Speech, Narrator's Representation of Voice, Narrator's Representation of Speech Acts, Indirect Speech, Free Indirect Speech, (Free)Direct Speech	Narrator's Representation of Speech Acts, Indirect Speech, Free Indirect Speech, (Free)Direct Speech	Narrator's Presentation of Speech Acts, Indirect Speech, Free Indirect Speech, (Free)Direct Speech
Presentation of Writing		Narrator's Report of Writing, Representation of Writing, Narrator's Representation of Writing Acts, Indirect Writing, Free Indirect Writing, (Free)Direct Writing	Narrator's Representation of Writing Acts, Indirect Writing, Free Indirect Writing, (Free)Direct Writing	Narrator's Presentation of Writing Acts, Indirect Writing, Free Indirect Writing, (Free)Direct Writing
Presentation of Thought	Direct Thought, Free Direct Thought, Indirect Thought, Free Indirect Thought, Narrator's Representation of Thought Acts	Narrator's Report of Thought, Internal Narration, Narrator's Representation of Thought Acts, Indirect Thought, Free Indirect Thought, (Free)Direct Thought	Narrator's Representation of Thought, Narrator's Representation of Thought Acts, Indirect Thought, Free Indirect Thought, (Free)Direct Thought	Narrator's Presentation of Thought Acts, Indirect Thought, Free Indirect Thought, (Free)Direct Thought

TABLE 2.2 *Functional comparison of presented discourse analytical models*

	Designation of model			
Function of presented discourse	**Leech and Short (1981)**	**Semino and Short (2004)**	**Short (2007)**	**Short (2012)**
Faithfulness to original discourse	Direct Speech, Free Direct Speech, Direct Thought, Free Direct Thought	(Free)Direct Speech, (Free)-Direct Writing, (Free)Direct Thought	(Free)Direct Speech, (Free) Direct Writing, Free Indirect Speech, Free Indirect Writing	
Focus on message rather than on wording	Indirect Speech, Indirect Thought	Indirect Speech, Indirect Writing, Indirect Thought		
Recreating the feel of the original / Blurring the line between narrator and characters	Free Indirect Speech, Free Indirect Thought	Free Indirect Speech, Free Indirect Writing, Free Indirect Thought		
Summarizing original discourse	Narrator's Representation of Speech Acts, Narrator's Representation of Thought Acts	Narrator's Representation of Speech Acts, Narrator's Representation of Writing Acts, Narrator's Representation of Thought Acts	Narrator's Representation of Speech Acts, Narrator's Representation of Writing Acts, Indirect Speech, Indirect Writing	(Free)Direct Speech, Indirect Speech, Narrator's Presentation of Speech Acts, (Free)Direct Writing, Indirect Writing, Narrator's Presentation of Writing Acts

	Designation of model			
Function of presented discourse	**Leech and Short (1981)**	**Semino and Short (2004)**	**Short (2007)**	**Short (2012)**
Telling			Narrator's Representation of Voice, Narrator's Representation of Thought Acts, Indirect Thought, Narrator's Presentation of Thought	
Distancing			Narrator's Representation of Thought Acts, Indirect Thought, Narrator's Representation of Speech Acts, Narrator's Representation of Writing Acts, Narrator's Representation of Voice, Narrator's Representation of Writing	
Introduction of presented discourse		Narrator's Report of Speech, Narrator's Report of Writing, Narrator's Report of Thought		
Minimal indication of discourse activity		Narrator's Representation of Voice, Narrator's Representation of Writing, Internal Narration		

TABLE 2.2 *Functional comparison of presented discourse analytical models* (cont.)

	Designation of model			
Function of presented discourse	**Leech and Short (1981)**	**Semino and Short (2004)**	**Short (2007)**	**Short (2012)**
Vividness, Drama, Showing			(Free)Direct Speech, (Free)-Direct Writing, (Free)Direct Thought, Free Indirect Speech, Free Indirect Writing, Free Indirect Thought	

Leech and Short's (1981) Model

The original study by Leech and Short (1981) and the subsequent revised edition of it (Leech and Short 2007), introduced a model of speech and thought presentation for fiction. However, the analytical categories they developed have since been proven to apply to non-fiction as well.

Below I include a complete list of the categories of speech and thought presentation found in Leech and Short (2007: 255–281). For each category I supply an example and give a description of the prototypical effect associated with that form of discourse presentation. My examples come from popular science and illustrate the model's applicability to non-fiction.

– Direct Speech/Direct Thought [DS/DT]—The original utterance or thought presented as if it were verbatim and introduced by a reporting clause (in blue).

Example of Direct Speech:

> Joel Garreau, author of *Radical Evolution*, said, "If a self-assembler ever does become possible, that's going to be one of history's 'holy s—!' moments. Then you are really talking about changing the world into something we've never recognized before."
>
> KAKU 2011: 205

Effects produced: Focus on the original speech or thought situation and faithfulness of representation. Emphasis on the originator of the speech/thought (Leech and Short 2007: 256–257).

- Free Direct Speech/Free Direct Thought [FDS/FDT]—Original discourse presented as if it were verbatim but without the reporting clause and often without the quotation marks. In the example below, Kean (2011: 140) puts Free Direct Speech in italics in order to indicate a shift from the narrator's voice to the collective voice of the scientific community.

Example of Free Direct Speech:

> But all his training also told Rous the idea was ridiculous—and his colleagues seconded that. *Contagious cancer, Dr. Rous? How on earth could a virus cause cancer?*

Effects produced: Focus on the character's voice without "the narrator as an intermediary" (Leech and Short 2007: 258).

- Indirect Speech/Indirect Thought [IS/IT]—Reformulation of an original utterance or thought that contains a reporting clause.

Example of Indirect Thought:

> Darwin realized that if all the offspring of plants or animals survived, population numbers would soon pass beyond the limits of what the environment could sustain.
>
> COEN 2012: 25

Effects produced: More complete integration into a narrative compared with DS/DT; focus on the message rather than on the exact words (Leech and Short 2007: 256–257).

- Free Indirect Speech/Free Indirect Thought [FIS/FIT]—Indirect Speech/Indirect Thought that is presented without a reporting clause. Free discourse reflects the narrative's perspective from the point of view of the character, which sometimes results in tense shifts.

Example of Free Indirect Speech:

> Far from rejoicing, the older scientist screwed up his brow and expressed his doubts that the nucleus contained any sort of special, non-proteinaceous substance. Miescher had made a mistake, surely.
>
> KEAN 2011: 20

Effects produced: While not a faithful reproduction of the original utterance or thought, it still has more power to refer to the feel of the original than Indirect Speech/Indirect Thought (Leech and Short 2007: 261).

– Narrative Report of Speech Acts/Thought Acts [NRSA/NRTA]—Summaries of utterances or thoughts.

Example of Narrator's Report of Thought Acts:

> On the same walk, Wilczek also came up with the idea for the "axion," a hypothetical low-mass cousin of the Higgs that is now a promising candidate for constituting the dark matter in the universe.
>
> CARROLL 2012: 169

Effects produced: Deemphasizing of the importance of the original utterance or thought in the new context. Emphasis on the fact that a speech or thought act took place not on the words or on the message (Leech and Short 2007: 259–260).

The value of Leech and Short's (1981/2007: 276) model is that it presents clear analytical categories for both speech and thought. As Short (2007: 226) observed, "It was the first attempt to distinguish systematically between speech and thought presentation." In this overview, I discuss the categories of speech and thought presentation together to highlight the fact that each category of speech has a counterpart in the presentation of thought and vice versa.

Leech and Short (1981) are also known for the introduction of the speech and thought presentation scales that arrange the categories of discourse presentation according to the degree of authorial control from the most controlled to the least controlled. The combined scale for speech and thought is represented below with the Narrative Report of Speech/Thought Acts being the categories most controlled by the author and Free Direct Speech/Thought being the least controlled:

> Narrative Report of Speech Acts/Narrative Report of Thought Acts
> Indirect Speech/Indirect Thought
> Free Indirect Speech/Free Indirect Thought
> Direct Speech/Direct Thought
> Free Direct Speech/Free Direct Thought

Leech and Short (2007: 276) use this arrangement to demonstrate that the modes of speech and thought presentation represent a continuum, with each category responsible for different effects on the reader depending on how much control over presented voices or thoughts an author wants to exercise.

Leech and Short also mark what they call "the norm" for speech and thought presentation. Thus the norm for speech presentation in fiction is Direct Speech, and the norm for thought presentation is Indirect Thought. These categories are chosen as the norms because each of them reflects presented discourse in the form it is directly expressed to the addressee (Leech and Short 2007: 276). In other words, if you are listening to someone talk, you are experiencing Direct Speech. And since thinking is an internal, unvoiced process, the only way it is accessible to others is via its indirect form. The norms, therefore, reflect the reality of typical interactions.

Semino and Short's (2004) Model

Semino and Short (2004) apply the model described above to a group of fiction and non-fiction texts. In the process, they demonstrate the need to expand the model by supplementing it with several new categories of speech/thought presentation and a whole new category of presented discourse—Presentation of Writing.

One of the unique features of their study is that it utilizes a corpus approach to test the Leech and Short model as it applies to written narratives beyond literary fiction. A corpus analysis is an investigation of linguistic patterns in a sample of naturally-occurring texts. In case of the Semino and Short study, the texts analyzed represented prose fiction, newspapers, biographies and autobiographies.

The new categories of discourse presentation introduced by Semino and Short demonstrate the authors' focus on creating a model that would account for all the possible instances of discourse presentation in their texts (Semino and Short 2004: 43). This is one of the strengths of their approach. With accuracy being the primary concern, Semino and Short pay less attention to the possible rhetorical effects associated with each of the categories they introduce compared to Leech and Short (1981), Semino (2004) and Short (2007, 2012).

The complete list of the categories used by Semino and Short is presented below. I am now going to demonstrate how the Semino and Short (2004) model applies to the data I compiled; I will provide a critical analysis and suggest adaptations for the model afterwards. Where possible, I will introduce examples from my corpus; however, for the categories where my data does not supply clear illustrations, I will quote from the Semino and Short (2004) corpus. As with the Leech and Short model, I will introduce the corresponding categories of speech, thought, and writing together.

– Narrator's Report of Speech/Thought/Writing [NRS/NRT/NRW]—clausal (as in reporting clause) or non-clausal "reporting signals" that introduce presented discourse (Semino and Short 2004: 35–39).

This category is broader than the traditional reporting clause as found in the examples of the Leech and Short model. It covers all possible ways of discourse introduction including passive constructions, noun phrases, etc.

Example from my corpus (Narrator's Report of Speech is in green, Indirect Speech that it introduces is in blue):

> To prove that there must be an infinite number of primes, Euclid started by asking whether, on the contrary, it was possible that there were in fact a finite number of primes.
>
> DU SAUTOY 2011: 32–33

This example includes a type of Narrator's Report of Speech that Semino and Short (2004: 37) identify as "'complex preposition'"—"To prove that there must be an infinite number of primes." Narrator's Report of Speech/Thought/Writing is not itself part of presented discourse but is used as an introductory element only.

– Narrator's Representation of Voice/Internal Narration/Narrator's Representation of Writing [NV/NI/NW]—minimal indications of verbal, thought, or writing activity without describing specific speech acts in any kind of detail (Semino and Short 2004: 43–45).

The functions assigned to this category by Semino and Short (2004: 44) are very close to those Leech and Short (1981/2007) attribute to Narrator's Representation of Speech/Thought Acts. In fact, according to Semino and Short (2004: 43–48), their categories differ from Leech and Short's only by the absence of message that is summarized. In other words, they are less detailed, minimalist versions of Leech and Short's Narrator's Report of Speech Acts and Narrator's Report of Thought Acts. There is no clear example of Narrator's Representation of Voice or Narrator's Representation of Writing in my corpus; the closest phenomena I observed would be classified as reporting clauses, for which Semino and Short (2004) have a separate category (see above). An example of Narrator's Representation of Voice from the Semino and Short (2004: 43) corpus is "She talked on." The example closest to this one from my corpus might be, "Robert Oppenheimer, who was chairing the proceedings, coaxed him into resuming his talk …" (Carroll 2012: 156). The relative clause in blue, I suggest, approximates the category of Narrator's Representation of Voice. For Internal Narration (NI)—the equivalent of Narrator's Representation of Voice in presentation of thought—Semino and Short (2004: 46), propose the following definition: "the presentation of [a] character's internal states" in the absence of a "specific thought act." Again, the example from my corpus can only approximate these requirements: "But all his training also told Rous the idea was ridiculous" (Kean 2012: 139). Semino and Short (2004: 46) supply the following example: "For a moment she didn't know where she was."

The difference between Narrator's Report of Speech/Thought/Writing and Narrator's Representation of Voice/Internal Narration/Narrator's Representation of Writing, as I interpret it, is that the first categories serve as introductory elements only and do not necessarily contain explicit indication of speech/thought/writing acts within themselves (as long prepositions demonstrate). Their purpose is to frame not to name the stretch of presented discourse that follows. The second categories are more concerned with an indication of a speech/thought/writing activity while they do not focus on the details of presented discourse.

– Narrator's Representation of Speech/Thought/Writing Acts [NRSA/NRTA/NRWA]—summaries of utterances, thoughts, or writing.

Semino and Short follow Leech and Short's definition and description of the functions for these types of presented discourse, emphasizing the summarizing properties. However, Semino and Short seem to have abandoned Leech and Short's (1981/2007: 26) notion that Narrator's Report of Speech Acts is used "for summarizing relatively unimportant stretches of conversation." Semino and Short (2004), in general, devote more attention to Narrator's Report of Speech Acts, Narrator's Report of Thought Acts and similar categories (see Narrator's Representation of Voice, Internal Narration, and Narrator's Representation of Writing along with some Narrator's Report of Speech, Narrator's Report of Thought, and Narrator's Report of Writing) and even introduce a subcategory—Narrator's Representation of Speech/Thought/Writing Acts with Topic (NRSAp, NRTAp, NRWAp). This new category covers Narrator's Representation of Speech/Thought/Writing Acts that include more detail than Narrator's Representation of Speech Acts and Narrator's Representation of Thought Acts from Leech and Short (1981). Semino and Short (2004: 52–53) define Narrator's Representation of Speech Acts with topic as the type of NRSA that "spells out the speech act that the original speaker is supposed to have performed …, and then goes on to provide details of the content of the utterance." Effectively, there are two parts in such Narrator's Representation of Speech/Thought/Writing Acts: one that "spells out the speech [thought/writing] act" and another one that introduces details. Semino and Short (2004: 53) acknowledge the similarity of Narrator's Representation of Speech Acts with topic to Indirect Speech and refer to other analysts who have classified such occurrences as "'condensed'" Indirect Speech (see, for example, Waugh 1995 cited in Semino and Short [2004: 53]). At the same time, Semino and Short insist on the NRSAp category. The following two examples represent Narrator's Representation of Speech Acts and Narrator's Representation of Speech Acts with topic. NRSA: "Penzias and Wilson called Dicke, who quickly confirmed that they had unintentionally tapped into the reverberation of the big bang" (Greene 2011: 40). NRSAp (the part that "spells out the speech act" is in green; the details are in blue):

> By that May, when Wheeler visited Bohr in Copenhagen and discussed Everett's ideas, the reception was icy. Bohr and his followers had spent decades refining their view of quantum mechanics. To them, the questions Everett raised, and the outlandish ways in which he thought they should be addressed were of little merit.
>
> GREENE 2011: 190

The Semino and Short examples of Narrator's Representation of Speech Acts with topic do not include those instances where the topic is detailed in a separate sentence, as in my example. I explain the difference by the fact that the NRSAp that Semino and Short (2004) observed were mostly found in the press section of their corpus and were used primarily, as Semino and Short (2004: 76–77) note, for their space-saving properties. I suggest that in popular science books, Narrator's Representation of Speech Acts both dramatize and summarize, so the length of NRSAp is not surprising.

– Indirect Speech/Thought/Writing [IS/IT/IW]—presenting "the contents of utterances [thoughts/writing] without normally claiming to reproduce the original wording" (Semino and Short 2004: 79).

Semino and Short (2004: 81) note that "IS [Indirect Speech] always involves a reported clause, which is typically introduced by a reporting clause containing a verb indicating speech activity." The same is true for Indirect Thought; however, the reporting verb is more likely to be a verb of cognition. At the same time, "the NRTs [Narrator's Reports of Thought] introducing IT [Indirect Thought] sometimes contain reporting verbs of speech, which are used to indicate silent self-address" (Semino and Short 2004: 129). Consider the following example of Indirect Speech from my corpus (reporting clause is in green, Indirect Speech is in blue):

> Two thousand years later, the physiologist Leonard Hill argued in the 1920s that they [colds] were caused by walking outside in the morning, from warm to cold air.
>
> ZIMMER 2011: 10

This is a prototypical example of Indirect Speech, where the reporting clause precedes the reported utterance. In my corpus, there are some utterances (but not thoughts) that are interrupted by a reporting clause. For example (reporting clause is in green, Indirect Speech is in blue):

> If something like water is heated, so that it evaporates and turns into a gas, the same corpuscles would still be there, said Boyle, but the gas occupies more space than the liquid had done.
>
> BYNUM 2012: 85

Such examples, though formally Indirect Speech, create an effect similar to that of Direct Speech, where authorial control is minimal. This effect I suggest, is different from the one produced by the prototypical variety of Indirect Speech. Semino and Short (2004) do not analyze examples of this kind. It is possible that their focus on reporting the types, not functions, prevented them from devoting attention to this slight variation.

– Free Indirect Speech/Thought/Writing [FIS/FIT/FIW]—Indirect Speech, Indirect Thought, or Indirect Writing that is presented without a reporting clause or any other marker of "narratorial control" (Semino and Short 2004: 86).

In this definition, Semino and Short (2004) admit that they follow Leech and Short (1981) and that other analysts regard Free Indirect Speech and Free Indirect Thought in slightly different terms, requiring only that there be no reporting clause (that is no grammatical dependence) but allowing for other markers of a narrator's presence, for example, quotation marks. Even though Free Indirect Speech, Free Indirect Thought, and Free Indirect Writing appear free of the narrator's control, these are still indirect forms of discourse, and their freeness, as Semino and Short (2004: 83) argue, may produce confusion, as it obscures the identity of the speaker and requires the reader to pay extra attention to determine to whom the discourse belongs. Here is an example of Free Indirect Speech from my corpus (FIS is in blue):

> Far from rejoicing, the older scientist screwed up his brow and expressed his doubts that the nucleus contained any sort of special, non-proteinaceous substance. Miescher had made a mistake, surely. Miescher protested, but Hoppe-Seyler insisted on repeating the young man's experiments—step by step, bandage by bandage—before allowing him to publish.
>
> KEAN 2012: 20–21

The tense shift from the simple past to the past perfect indicates a change in the narrative focus and introduces some degree of immediacy and surprise that the reader is to associate with Hoppe-Seyler. In this example, the identity of the speaker is less obscured as in some; however, it is still possible to confuse Free Indirect Speech with narration especially in the second selection (Semino and Short [2004: 83] also note a similar problem); however, the tense shift helps to distinguish Free Indirect Speech from other forms of discourse presentation and narration.

– (Free)Direct Speech/Thought/Writing [(F)DS/(F)DT/(F)DW]—original discourse presented as if it were verbatim.

Quotation marks and reporting clauses may be present but are not required. The focus is on the fidelity of the presentation of the original. Semino and Short (2004: 88) express their preference for not separating the free forms from their direct counterparts (In this they admit they contradict Leech and Short.). However, Semino and Short (2004: 88) also admit that they used the two separate tags (one for free and one for direct forms) when the distinction was functionally significant. Such instances are not given much attention, and Semino and Short (2004) clearly prefer the combined category, to which they assign the functions usually associated with DS. Dramatization and character creation are the main roles of (Free)Direct Speech/Thought/Writing. For Semino and Short, the presence or absence of quotation marks and/or reporting clauses does not affect how the reader perceives the discourse presented. They argue that most instances of dialogue omit the formal marks of direct discourse, yet do not aim at a different effect by doing so. Here is an example of Direct Speech from my texts (it includes a reporting clause in blue): "Rous himself later admitted, 'I used to quake in the night for fear that I had made an error'" (Kean 2012: 141). Most instances of (Free)Direct Speech in the popu-

TABLE 2.3 *Speech, thought, and writing presentation scales based on Semino and Short (2004)*

Presentation of Speech	[N]	NV	NRSA	IS	FIS	DS(FDS)
Presentation of Writing	[N]	NW	NRWA	IW	FIW	DW(FDW)
Presentation of Thought	[N]	NI	NRTA	IT	FIT	DT(FDT)

lar science books I analyzed include quotation marks but do not use reporting clauses—thus representing a mixture of Direct Speech and Free Direct Speech—which tends to argue for Semino and Short's (2004) combined category.

As a result of the additions that Semino and Short introduce, the different categories can be arranged on a cline of directness shown in Table 2.3.

Semino and Short use [N] to represent narration—a mode of writing that involves no discourse presentation. It is outside the discourse presentation scales (as the square brackets indicate) just as are Narrator's Report of Speech, Narrator's Report of Thought, and Narrator's Report of Writing. Even though Narrator's Report of Speech/Thought/Writing are connected more closely to discourse presentation than narration, these categories themselves only introduce presented discourse, and Semino and Short do not include them as part of the scales.

Overall, Semino and Short's (2004) contribution is the introduction of Presentation of Writing as a category of presented discourse alongside speech and thought presentation. They also apply a more detail-oriented approach to the analysis of the already existing categories, which leads them to the distinction between Narrator's Representation of Voice, Narrator's Representation of Writing, and Internal Narration on the one side and Narrator's Representation of Speech Acts, Narrator's Representation of Thought Acts, and Narrator's Representation of Writing Acts on the other. In general, however, the attempt to account for all the possible forms of discourse presentation in fiction and nonfiction forced Semino and Short to spend less time analyzing the functions of the discourse presentation categories. This is something that Short (2007, 2012) takes upon himself to remedy. In the following subsection, I will discuss his commentary on the Semino and Short (2004) model.

Short's (2007, 2012) Commentary on the Semino and Short (2004) and Leech and Short (1981) Models

Firstly, if Semino and Short (2004) focused primarily on the frequency information, Short (2007) devotes more attention to the functions of speech, thought, and writing presentation. In my opinion, that strengthens the 2004 model. On the other hand, if the 2004 study analyzed a sample of both fiction and non-fiction, Short (2007) reverts to giving examples of, and drawing conclusions from, fiction texts alone.

Short (2007: 230) ascribes the following functions to the presentation of speech and writing: *Faithfulness* (with Free Indirect Speech / Free Indirect Writing and (Free)Direct Speech / (Free)Direct Writing being the best options for its fulfillment), *Summarizing* (with Indirect Speech / Indirect Writing and Narrator's Representation of Speech Acts / Narrator's Representation of Writing Acts supplying prototypical examples), *Distancing* (best fulfilled by Narrator's Representation of Speech Acts /Narrator's Representation of Writing Acts and Narrator's Representation of Voice / Narrator's Representation of Writing), *Telling* (with Narrator's Representation of Voice being the only prototypical option), and *Vividness, Drama*, and *Showing* (best achieved by the use of Free Indirect Speech / Free Indirect Writing and (Free)Direct Speech / (Free)Direct Writing).

Short (2007: 230–231) notes that thought presentation is markedly different from the speech/writing category in its functions. Short (2007: 230–231) suggests removing from thought presentation the functions of summarizing and faithfulness. Short (2007: 230) argues that "there is something of an issue, of course, concerning to what extent, outside fiction, our thoughts come to us in linguistic form at all." Therefore, thought presentation cannot be assessed for accuracy. His argument about excluding the summarizing function from the effects of thought presentation is as follows, "There is no possibility of assuming a fictional 'original' which can be approximated to, more, or less, in the presentation of thought" (Short 2007: 231). Instead, Short (2007: 321) proposes to focus on the "effects relating to the extent of narrator 'interference'" when it comes to thought presentation.

The effects that could be achieved via thought presentation, as Short (2007: 231) suggests, can be combined into two opposing groups. The first functional group is a result of more intensive interference from a narrator and consists of *Distancing* and *Telling* (achieved through the use of Narrator's Representation of Thought Acts and Indirect Thought). The second functional group depends of the lesser degree of a narrator's involvement and consists of *Vividness, Drama*, and *Showing* (to be achieved via Free Indirect Thought, Direct Thought, and Free Direct Thought). For a comparison of functions assigned to various presented discourse categories, see Table 2.2 above.

Another important contribution of Short (2007) is the reevaluation of the Internal Narration (NI) category. If Semino and Short (2004) placed it inside the discourse presentation scale, Short (2007: 234), following Toolan's (2001) suggestion, argues for placing it outside the scale. Short (2007: 234) explains that "cases of NI are the statements that the narrator makes about the inner world of his or her characters" not statement made by the characters themselves.

Short (2007: 234) suggests a new category of Narrator's Presentation of Thought (NT). Narrator's Presentation of Thought is used to create the effect of telling; it is an example of a narrator's interference and control. As Short (2007: 235) notes, "NT is used infrequently in novels" because the authors usually prefer to rely on discourse presentation that shows how the characters feel rather than telling the readers about it. Short (2007: 235) also suggests that the contrast between Free Indirect Thought and Narrator's Presentation of Thought can be used by authors for strategic purposes. For example Short (2007: 235) finds that Narrator's Presentation of Thought is often employed in detective stories to give the reader clues as to who might be the murderer. Narrator's Presentation of Thought, according to Short (2007: 234–235), can also be used to place the reader inside the mind of a character but without triggering an emotional attachment. Short (2007: 234–235) argues that Narrator's Presentation of Thought produces this effect because unlike Free Indirect Speech it does not let the reader be "privy to what [the character] is thinking;" Narrator's Presentation of Thought alerts the reader only to the fact that the character is thinking of a specific subject. Consider Short's (2007: 234–235) example of Narrator's Presentation of Thought from a novel by Iain Banks, which uses second-person narration for the part of the story:

> The reader is positioned in the mind of a serial murderer as he kills one of his victims … it soon becomes clear by inference that he is experimenting with ways of making his victim suffer before he finally dies.
>
> You let him slump back again so that he's sitting against the chicken-wire gate and when his eyes start to open you pull his head forward by the hair and cosh him again. He falls to the side. You put the plastic ties in your pocket. *You're thinking.* The foxhounds continue barking and yelping.
>
> BANKS 56–57, SHORT's emphasis

On the thought presentation scale, Narrator's Presentation of Thought replaces Internal Narration (NI), which is now included in narration, that is, not part of presented discourse at all. Thus Narrator's Presentation of Thought becomes the most narrator-controlled form of thought presentation (Short 2007: 235).

Short (2012) is a continuation of, and a comment on, both Leech and Short (1981/2007) and Semino and Short (2004). In this study, Short (2012) brings up two important aspects that, as he demonstrates, he has missed or misinterpreted in earlier works. Both of the issues are connected with the idea of faithfulness in discourse presentation, which Short (2012) defends against the claims of other researchers. To make his argument, Short (2012: 19) proposes a clear distinction between report and representation of discourse, where report assumes absolute accuracy, and representation implies "a *mismatch* between the lexis, deixis and grammar of the anterior and posterior discourses" (Short 2012: 19). In other words, a representation can change the perspective of an utterance to indicate that it is being reported rather than experienced. Presentation, in this framework, may refer to any kind of posterior discourse, either report or representation (Short 2012: 19). This distinction leads Short (2012) to reconsider some of the terminology used in earlier studies. Thus he suggests replacing the term *Representation* "(and so the acronym 'R')" found in the earlier category names (NRSA—Narrator's Representation of Speech Acts, for example) with *Presentation* "(and so the acronym 'P')" to form NPSA—Narrator's Presentation of Speech Acts, for example (Short 2012: 21). The substitution, Short (2012) suggests, will reflect the nature of the discourse presented more accurately.

Another suggestion that Short (2012) makes is that the idea of faithfulness does not always apply even to such prototypically "faithful" categories of presented discourse as Direct Speech. As Short (2012: 23–24) proposes, this is because Direct Speech (and other forms of direct as well as indirect discourse) can be used to create summaries. In his earlier works, for example, Leech and Short (1981) and Semino and Short (2004), he argued that only indirect discourse fulfills the summarizing function. In the later study, he allows the possibility that direct discourse too may have summarizing properties (Short 2007: 230); however, the idea is expressed rather tentatively, and the summarizing functions are described as "prototypically associated with NRSA" (Short 2007: 230). Later, Short (2012) fully develops the idea that any discourse category can be used to summarize.

Short (2012: 23–24, 29) proposes two types and one subtype of summaries created through discourse presentation: proposition-domain summaries and discourse-domain summaries (with quotative summaries as a subtype).

A proposition-domain summary is a kind of summary that deals with one proposition. This is what was understood as summary in Leech and Short (1981) and Semino and Short (2004). A proposition-domain summary is represented by Narrator's Presentation of Speech Acts and Narrator's Presentation of Writing Acts (Short 2012: 23–24). A discourse-domain summary deals with a large piece of discourse that contains several propositions. Any category of discourse presentation (except for presentation of thought) can produce a discourse-domain summary. A quotative summary is a summary "which quotes representative parts" of discourse to create a discourse-domain summary (Short 2012: 29).

TABLE 2.4 *Discourse presentation scales reflecting the summarizing function*

Presentation of Speech	NPVs [Narrator's Presentation of Voice, formerly NV]	NRSAs	ISs	FISs	DSs
Presentation of Writing	NPWs [Narrator's Presentation of Writing, formerly NW]	NRWAs	IWs	FIWs	DWs

As a result of the new approach to the summarizing functions of presented discourse, Short (2012) introduces new scales for speech and writing presentation of discourse-domain summaries. These scales are similar to the discourse presentation scales for speech and writing except that each category receives a subscript "s" to indicate the summarizing properties. Table 2.4 is a representation of the scales from Short (2012: 28). Thought presentation is excluded from the scales since Short (2012) does not regard presentation of thought as possessing any summarizing properties.

As I have demonstrated, Short (2007) makes an important addition to the Semino and Short (2004) study by focusing on the functions of presented discourse and supplying clear prototypical categories associated with each possible effect. The functional distinction between speech/writing and thought presentation, in my opinion, becomes more prominent in this later work than it was in Semino and Short (2004). Short (2012: 22) makes this difference even clearer when he writes that "Thought presentation is *not* the presentation of communication between people but the presentation of someone's inner world" (emphasis in the original). The implication of this conclusion is significant and is prominent in Short (2012), where thought presentation is excluded from the discussion of summaries more explicitly than it is in Short (2007). Functional approach to discourse presentation also makes it possible for Short (2012) to suggest changes in terminology, which reflect the roles of discourse presentation categories more accurately.

If the three discourse presentation models discussed here were combined, the scales originally proposed by Leech and Short (1981) would look as follows (square brackets indicate categories outside the scales but related to discourse presentation—Narrator's Report of Speech/Writing/Thought—or contrasted with them—narration and Internal Narration). See Tables 2.5.a, 2.5.b, 2.5.c.

TABLE 2.5.A *Combined speech presentation scale*

[N]	[NRS]	NPV	NPSA	IS	FIS	(F)DS

TABLE 2.5.B *Combined writing presentation scale*

[N]	[NRW]	NPW	NPWA	IW	FIW	(F)DW

TABLE 2.5.C *Combined thought presentation scale*

[N, including NI]	[NRT]	NPT	NPTA	IT	FIT	(F)DT

Fictionality: Questions about Definition

The idea of fictionality is firmly associated with literary fiction (novels, short stories, etc.); on the other hand, studies in non-fiction (biographies, journalistic accounts, speeches, etc.) have also demonstrated a connection with fictionality for texts that are based on real-life events. The two approaches to fictionality—in fiction and in non-fiction—produce two different definitions of the phenomenon, which I will discuss in this section. I should say that I did not expect to find any significant manifestations of fictionality in popular science, but in the process of analyzing presented discourse of scientists, I was alerted to an unusually consistent presence of fictionalizing mechanisms.

The way presented discourse in popular science narratives is used to turn the stories from dry facts to engaging fictions does not correspond directly with any of the definitions of fictionality in either literary fictions or non-fiction texts. Rather, it blends the two established approaches. The placement of the fictionalizing elements within the presented voices is significant. It shows that the scientists, as introduced to the readers of popular science, are not just information-delivery mechanisms but carefully crafted characters whose goal is to appeal to the reader and to project a likable and exciting image of the scientific community.

Before I go further into the discussion of fictionality in popular science and its implications, it is necessary to understand what fictionality means in literary fiction and in non-fiction. Fludernik's (1996) definition of fictionality exemplifies the literary fiction approach to the phenomenon: literary fictionality is seen as "the subjective experience of imaginary human beings in an imaginary human space" (Fludernik 1996: 39). Fludernik emphasizes "the subjective experience," saying that literary fictions "set out to represent" the human experience and create an "evocation of 'real-life' experience" (Fludernik 1996: 41, 12). According to this understanding of fictionality, to be considered fiction, a text must project the emotional experiences of its characters since emotional reactions are more unique and subjective than physical responses. Dramatization not only of external events but also of thoughts and feelings is essential to literary fiction.

This approach to fictionality has become synonymous with literature, and the focus on the subjective experiences of the characters in a story has become a defining feature of narrativity (see, for example, Toolan 2001 and Herman 2009). Thus Toolan (2001: 8–9) argues that "the subjective experience" alone is enough to establish a text as a narrative even if the structural elements of a proper story are lacking. Toolan (2001: 8) stresses the value of the characters' experiences to such an extent that he calls them "experiencers." For him, it is the experiences of the characters and their interpretations of these experiences that connect events in a meaningful way thus creating a story (Toolan 2001: 7–9). Herman (2009: 139–140) also lists the characters' experiences and consciousness as some of the most notable features of narrativity in either fiction or non-fiction, though he chooses to illustrate them with fictional literary narratives primarily.

The view of fictionality as exemplified by Fludernik's (1996) definition with its stress on the experiences of the characters is so pervasive that even those who do not explicitly incorporate her approach can be seen as being influenced by it. For example, the idea of the transfer of an experience from a character to a reader is a key feature of a literary narrative according to Caracciolo (2013). His study examines phenomenological metaphors—a specific way in which "the subjective experience of imaginary human beings" of Fludernik's definition can be transferred to the reader.

An important aspect of Fludernik's definition—the qualification "imaginary" assigned to the characters and their surroundings—is taken for granted when discussing literary fictions. However, it comes to the forefront when we try to apply the definition to non-fiction. How can "imaginary human beings" or "imaginary human space" fit into a non-fiction narrative, especially one concerned with science? The immediate answer is they cannot; this is, essentially, what separates non-fiction from fiction. If I were to apply Fludernik's definition to popular science, there would not be any signs of fictionality. The popular science books I analyzed would pass the non-fiction test. Even the text segments that focus on the experiences of the characters would still lack the "imaginary" aspect. This is why I have chosen to illustrate this definition of fictionality with an example not from the texts I investigated but from a scientifically-themed novel *Intuition* by Allegra Goodman (2006). The novel belongs to the genre of Laboratory Literature or lab lit. Rohn (2005, 2010a) defines lab lit as realistic fiction about scientists set in a world identifiable as cognate with our own (as opposed to the speculative settings found within science fiction). Goodman's (2006) work deals with a group of scientists working in a cancer research laboratory who come to experience the exaltation of discovery and the consequences of fraud. In the process, they reexamine the nature of their personal and professional relationships.

Example 2.1 uses presentation of thought of two scientists to share their experiences in the laboratory with the reader:

2.1 Marion and Feng looked at each other. After repeated failure, could one of Cliff's viral variations actually have some effect? What had changed here? What had Cliff done? The variation of the virus was R-7, Cliff had scrawled a note on the blue index card labeling this cage of mice. But he's never gotten R-7 to work effectively in live animals before. Were these three mice significant? Or were they outliers of some kind—tainted by some other condition? This was the difficulty with animal research: so many different things could go wrong.

GOODMAN 2006: 27

This example contains Free Indirect Thought (FIT) of scientists (The use of the past tense and the past perfect suggests Indirect rather than Direct Thought.). The author uses FIT to express the emotions of puzzlement and hesitation. Goodman (2006: 27) never indicates that these questions are voiced, and the reader is to presume that he is given access to the inner worlds of the characters. The choice of Free Indirect Thought in this passage is very likely not accidental since, as Semino and Short (2004: 123–124) show, this is the most frequently used form of thought presentation in fiction. Free Indirect Thought is used most often to deliver thoughts in a "dramatic and immediate way" that allows the reader access to "the consciousness of characters" and creates "closeness and empathy towards characters" (Semino and Short 2004: 123–124). The reader is allowed to share in these unspoken emotions which reflect the subjective experience of the scientists.

Thought presentation is a common way to let the reader co-experience what Fludernik calls "an imaginary human space," which is inhabited by the characters. Because it is only in fiction that a reader *knows* what other people are thinking and because it is only in fiction that thoughts come in well-formed words and grammatically correct sentences, Cohn (1990), Short (2007), and

Dawson (2015) are able to suggest that any instance of thought presentation immediately signals fictionality. However, my findings suggest that the relationship between thought presentation and fictionality is more complex in popular science. Even if the qualification "imaginary" is disregarded, and popular science is assessed on the basis of the presentation of subjective experiences alone, not all instances of thought presentation would automatically point to the presence of fictionality because in popular science not all (in fact, hardly any) instances of thought presentation evoke exclusively subjective experiences. As I will demonstrate in chapter 3 and explicate in chapter 5, thought presentation is rarely used in the way that Goodman's novel uses it. And the category of Free Indirect Thought—the most dramatic and immediate of all the forms of thought presentation—is entirely absent from popular science stories of discovery. The reader of popular science is much more likely to encounter thought presentation similar to the instance in example 2.2.

2.2 Alvarez, a brilliant physicist with a penchant for puzzles, hit on an idea: Use muons coming from the skies in the form of cosmic rays to peek inside the rock of Khafre's pyramid.

CARROLL 2012: 106

This is Direct Thought (DT) expressing a scientific hypothesis. This choice of the form for thought presentation is not accidental. It fits closely with the purpose of popular science texts and especially with the purpose of the discovery narrative from which this example is extracted. Semino and Short (2004: 118) note in contrasting Direct Thought with Free Indirect Thought that the former presents "highly conscious, deliberate thought" while the latter is more emotional. Deliberate and well-articulated thought expressed in 2.2 is a perfect way to express a scientific hypothesis that is expected to be tested. The reporting clause, "Alvarez ... hit on an idea" contains an element of surprise, yet it also sets up the instance of Direct Thought that follows as a product of previous deliberation on the subject.

Example 2.2 shows that in popular science narratives of discovery thought presentation is used for very different purposes than it is in fiction. Even though the author shares with the reader the thoughts of the scientist, they are not intended to evoke in the reader the emotional experience of coming up with a hypothesis and thus reveal an inner world. In fact, the author gives no indications of how Alvarez felt about his idea, whether, for example, he was excited or concerned that it might not work.

Even though no subjective experience is shared with the reader through the display of Alvarez's thoughts, some researchers (see, for example, Skov Nielsen et al. 2015a, b) would still regard this portion of the narrative as an instance of fictionality, and their justification is different from the idea that any instance of thought presentation automatically signals fictionality. Obviously, this kind of fictionality is different from the one identified by Fludernik (1996). This is fictionality as it has come to be recognized in non-fiction.

The attention to fictionality in non-fiction and the subsequent definition of the phenomenon is strongly connected with the idea of fictionality in science. Studies in philosophy of science have been advocating for a non-literary kind of fictionality as early as the beginning of the 20th century (see, for example, Vaihinger 1924). More recently, Arthur Fine (1993) has revived these ideas, which inspired a number of further explorations by philosophers of science (see, for example, Suarez 2009, Toon 2012, Barwich 2013). Fictionality, from this scientific point of view, is usually understood as "the role played by particular methods of model building such as abstractions, idealisations, and the employment of highly hypothetical entities" (Barwich 2013: 357–358). This view of fictionality focuses on an important aspect of the scientific process—hypothesizing. The aspects of imagination and experientiality present in Fludernik's definition are replaced with the discussion of the possibilities and probabilities.

Narratologists Skov Nielsen, Phelan, and Walsh (2015a, b) spelled out what such an approach might mean for fictionality in non-fiction and how fictionality might be recognized in a non-fiction text. I should note, however, that Skov Nielsen et al. (2015a, b) do not credit philosophy of science for this approach even though the similarity of their parameters for fictionality and those presented by the philosophers of science is clear. Skov Nielsen et al. (2015a: 62) suggest that fictionality can be expressed through "what-if projections, if-only regrets, thought experiments, and hypotheses of all kinds." In other words, any kind of hypothetical expression can be pointed to as fictionality if found in non-fiction. The experiential aspect is no longer vital; thus dramatization of either external events or internal states has no place in this view of fictionality.

Skov Nielsen et al. (2015a, b) stress that fictionality is "ubiquitous" (2015a: 62) and "extremely pervasive" (2015b: 110), yet at the same time it remains a phenomenon associated primarily with fiction. They assert that "apart from the work by literary critics on generic fiction, fictionality is almost completely unstudied and often unacknowledged" (Skov Nielsen et al. 2015a: 62). The point of Skov Nielsen's et al. (2015a, b) approach is to enable scholars and researchers to recognize and examine fictionality and its functions outside of the domain of literary fiction (Skov Nielsen et al. 2015a: 70). The desire to separate fictionality from fiction could explain the renunciation of the dramatizing and the emotional aspects from the definition.

On the other hand, by using a narrow definition of fictionality—the one that includes primarily hypothetical scenarios—Skov Nielsen et al. (2015a, b) limit the kinds of fictionality that could be observed in non-fiction. Such an approach, in my opinion, cuts out an important aspect of fictionality—that of recreation of reality through dramatization and emotion. I think this is an especially valuable aspect of fictionality for popular science because multiple linguistic studies as well as studies in public perception of science and studies in communication of science to the public suggest that lay readers rely heavily on an emotional connection with science and scientists that is introduced through popular science texts (see, for example, Calsamiglia 2003, Moirand 2003, Myers 2010, Laslo et al. 2011, Luzón 2013, Sackler 2014, Supper 2014). Therefore, I find it more productive not to exclude the emotional aspect of fictionality from non-fiction.

There are similarities between fictionality as it is expressed in both literary fiction and non-fiction. Philosophers of science (see, for example, Cartwright 1983, Toon 2012) have circulated the idea that fictionality in science is "analogous to literary … fiction" (Rouse 2009: 37). Semino and Short (2004) in their exploration of presented discourse in fiction and non-fiction concluded that the divides between fiction and non-fiction are less rigid than generally perceived. At the same time, the definition of fictionality proposed for one type of text does not necessarily recognize the manifestations of the same phenomenon in a different type of text. Thus the two separate definitions as they stand do not supply an explication for the elements of fictionality found in presented discourse of scientists in popular science texts.

The kind of fictionality connected with presented discourse is grounded in real events that are dramatized by the authors in order to introduce the experiences of the scientists. This is part of a long-standing tradition of theatricality in professional and popular science. For instance, Coppola (2016: 3) draws attention to the fact that in the eighteenth century, "there was no fundamental divide between the theatrical and the scientific, and the boundary between the stage and the laboratory broke down." He continues to suggest that as the findings of science became more readily available to the public, "natural philosophy became more theatrical … and more invested in the securing of mass assent through the deployment of convenient fictions and affective displays."

In modern popular science books, the events that are most often dramatized through presented discourse are conferences, debates, and conversations that took place in real life, but which are presented to the reader from the point of view of one character or another so that the reader can catch a glimpse of what it is like to be a scientist and to negotiate professional relationships. These events are not imaginary but reimagined to present science and scientists as emotionally relatable. Even hypotheses introduced in popular science narratives of discovery are usually expressed as thoughts, which suggests an attempt to make these inherently scientific portions of the stories more personal.

To examine popular science for fictionality, a different approach should be considered, one that is willing to regard imagination not in direct opposition to reality but as a mechanism that can enhance a reader's experience with an unfamiliar world.

My own definition of fictionality, which comes from the examination of presented discourse in popular science narratives of discovery, is as follows: I regard fictionality as *reimagining and reconstructing of real-life events in such a way that brings out subjective experiences of human actants*. I further propose that dramatization by means of emotionally-charged presented discourse is the way that popular science narratives of discovery fictionalize their subjects. The characters-scientists in these narratives shift the focus from the impersonal descriptions of science to the understanding of its human participants.

The Importance of Emotionality in Popular Science

It is common to associate emotionality and evocation of personal experience with literary fiction. For example, Fludernik's definition of fictionality introduced in the previous section is tied to these concepts, yet they are not often observed in non-fiction. Even though fictionality is not limited to a set of features that are present in literary fictions only, it is still somewhat unusual to regard emotionality as a key feature of non-fiction. After all, "emotional experience ... and the close connections, too, between emotional, ethical, and intellectual responses" that literary fiction provides are "a main" reason why we read fiction (Toolan 2011: 1). At the same time, all of these features are present in popular science as well; moreover, they are one of the reasons for the popularity of the genre. In fact, the emotional engagement of the public with science has become the major means of connection between scientists and lay people. For instance, Arthur Sackler (2014: 10), reporting the results of a colloquium on science communication in Washington, D.C., proposes that scientists would be better received by the public if they focused on such personal and emotional aspects "as letting people know why someone went into science."

The lay public values emotional engagement with science because for a long time it was the only means of connection between the experts and the non-professionals. Emotional engagement, I propose, has come to replace the active, hands-on involvement of the public common in the past centuries before the enforced separation discussed in chapter 1. The popularizers of the nineteenth and early twentieth century, guided by the dominant model approach, focused mainly on transmission of facts. The assumption that their audience, for the most part, was scientifically illiterate and lacked the tools to evaluate and connect with the knowledge transmitted did not allow the authors to explore the full potential of the public. However, knowledge of scientific facts is not the same as public engagement with or interest in science. For the scientists, knowledge is primarily a cognitive category, but for the public it can also be experiential and emotional, residing in the communicative dimension rather than entirely in the cognitive one. Myers (2003: 269) points out that the public's strength is in the ability to evaluate scientific discoveries for their possible benefit or detriment to society.

The assumption that knowledge is unilateral and concerns factual information exclusively is one more answer to the question why the separation between the scientific community and the public took place. I believe that the underlying cause of the gap is the fact that the public and the scientists rely on different types of knowledge. In addition, it is possible that each sees its preferred knowledge type as superior to and separate from the others.

Bloom's taxonomy of knowledge-types (see, for example, Bloom et al. 1956, Anderson and Krathwohl 2001, Krathwohl 2002) is usually seen as a standard in the field of knowledge acquisition. However, it is not the only model out there. Yang's (2003) holistic theory of knowledge stands out as it provides an update on the four commonly recognized knowledge types (factual, conceptual, procedural, and meta-cognitive) and demonstrates that they function not independently of each other but as a whole. Yang suggests the following facets of knowledge: explicit, implicit, and emancipatory. Explicit knowledge is the "cognitive component of knowledge that represents one's understanding about reality" (108). Implicit knowledge is "personal, content-specific familiarity" and "usually comes from … one's behavior, action, and accumulated experience" (109). Emancipatory knowledge is affective knowledge. It is "value-laden" and "is indicated by feeling and emotions people have in relation to the objects and situations around them" (109). Yang (2003: 108) argues that people learn "not only through mental correspondence but also by direct personal involvement [and] … through … emotions." He suggests that the three facets or types of knowledge are inseparable and that measuring just one of them provides false impressions about expertise or intelligence (111–113).

Yang's perception of knowledge helps explain the less than stellar performance of the lay public in surveys designed to check for knowledge of scientific facts (examples of which could be found in Myers 2003: 268 and Laslo et al. 2011: 848). It is plausible that the surveys are measuring explicit knowledge, while the public is more likely to possess implicit and emancipatory knowledge when it comes to science. Explicit knowledge, on the other hand, is the primary knowledge domain of the scientific community (as de Jong and Ferguson-Hessler 1996 show), and in trying to communicate with the public and to assess the public's understanding of science, the tools that the scientific community uses are possibly disregarding the other two knowledge facets. The preference for one knowledge type over another is a strong contributing factor to the dissociation between the scientists and the public (Pilkington 2016).

If the public connects with science and its practitioners on an emotional level, why shouldn't popularizations reflect this preference?

By the end of the twentieth century and beginning of the twenty-first, popularizations started to be recognized as points of interaction between the scientists and the public. In fact, interactivity became so prominent in popular science texts written in English as to allow Kranich (2016: 23) to list it among the features of English-language popularizations. With that, the way the public is seen by the scientific community has also began to shift. Today, the public is once again regarded by the scientists and the popularizers as a participant in scientific affairs (see, for example, Calsamiglia 2003, Laslo et al. 2011, Luzón 2013, Moirand 2003, Myers 2010, Supper 2014). However, the role of a lay person has changed from what it was in the past. Lay people today participate in science not so much by producing it but by experiencing it, evaluating it, and engaging with it emotionally. The public of the past used to engage with science in terms of explicit knowledge; the public of today relies more heavily on implicit and emancipatory knowledge of science. Recent linguistic studies support this conclusion.

Linguistic studies of popularization undertaken in the early twenty-first century show a clear shift in popular science towards the establishment of an emotional connection between the readers and the scientists. For example, Turney (2004a) and Caracciolo (2013) suggest certain metaphors can help the reader engage with science on an emotional level. Moirand (2003) suggests a more direct way to establish an emotional connection between the public and the scientific community—the use of presented discourse of scientists. Moirand (2003: 181) shows that expert voices and the way they are framed in a text create "a representation of the different attitudes, emotions and reactions of the speakers." Multiple voices can be used to express emotions within the scientific community and by doing so present scientists as more relatable to the public. Moirand (2003: 197) also suggests that in some cases the use of presented discourse of scientists shifts attention from the scientific facts to the emotional evaluations of the subjects discussed. Moirand's (2003) research establishes the connection between the reader and the presented voices of scientists as a basis for a model of popularization where the scientists and the public influence each other. Fu and Hyland (2014: 141) as well stress the importance of creating a connection between a lay reader and a professional scientist when they say that one of the roles of popular science is to allow "non-specialists to recover the voice of the scientist" (141). Moirand's (2003) findings suggest that this voice could be more emotional than expected of a non-fiction text.

I interpret Moirand's (2003) observation as evidence of dramatization and emotionality in popular science and so do Calsamiglia and Ferrero (2003: 169), who write, referring to Moirand (2000)—an earlier study of the same issue—that "... the very many voices involved in the problem ... dramatize the conflict." In other words, presented voices can contribute to the emotional aspect of fictionality. In the chapters that follow, I will examine how presented discourse of scientists is used to dramatize their professional experiences and in doing so establishes an emotional connection with the reader, not unlike a connection between readers of literary fiction and imaginary characters.

Dramatization, Emotionality and Professional Science

The use of dramatization and emotional connection between the scientific community and the reader may raise questions about the accuracy of the information introduced in a popular science text. At the same time, dramatization and emotionality are part of scientific research and discovery process. The fact that popular science narratives of discovery include these aspects makes them, on some level, more accurate in representing the work of the scientific community, mimicking the experiences involved.

Philosophers and sociologists of science agree that science is not all about facts; there is plenty of room for emotion (see, for example Gilbert and Mulkay 1984, Harré 1994). The focus on the results and facts prevalent in professional scientific publications, Harré (1994: 87) argues, is nothing more than "a smiling face" of science and has little to do with the actual process of discovery. Harré (1994: 90) suggests that "the harsh life of the scientific jungle reveal[s] itself" in "the microsociology of science," that is, the intermediate stage between "polished presentations" and "nascent research programmes."

Gilbert and Mulkay (1984) made a similar observation earlier when analyzing the language of the scientific community. What they describe as "empiricist discourse" corresponds to Harré's (1994) "polished presentations stage" and focuses on science exclusively. The intermediate stage, Gilbert and Mulkay (1984) suggest, manifests in "contingent discourse," which adds personal perspectives and incorporates issues that lie beyond the realm of science and sometimes conflict with the discussion of science. They stress that luck, chance, and interpersonal relationships are prominent contributors to scientific progress and label them "contingent factors" (Gilbert and Mulkay 1984: 92–95).

Popular science seizes these somewhat unscientific features of the discovery process and amplifies them in order to make scientists into relatable characters. In a way, popular science authors use fictionality to reintroduce the part of the discovery process that does not always extend beyond the laboratory. The authors accomplish this by using presented discourse. Consider example 2.3, which shows the final part of a narrative describing the discovery of the double helix structure of DNA:

2.3 Chargaff tried to relay this finding to Linus Pauling—Watson and Crick's main rival—while on a cruise, but Pauling, annoyed at having his holiday interrupted, blew Chargaff off. The cagier Watson and Crick heeded Chargaff (even though he thought them young fools), and from his insight they determined finally, that A pairs with T, C with G.

KEAN 2012: 100–101

As I will discuss at length in chapter 4, in this narrative Kean (2012) uses dialogue expressed primarily via Narrator's Presentation of Speech Acts (blue), and he also uses presentation of thought (green) to create this exchange. The interaction introduced in this example points to the importance of interpersonal relationships and chance encounters in discoveries. Kean (2012: 100–101) implies that had Pauling listened to Chargaff, he would have been the one to discover the structure of DNA. This is a suggestion that scientific truths are sometimes revealed by those who possess a kind of interpersonal cleverness rather than superior scientific intelligence. Professional scientific publications, according to Harré (1994) explicitly deny this path to discovery; popular science, on the other hand, brings it to the forefront.

Gilbert and Mulkay (1984: 91–95) suggest that even when scientists acknowledge and rely on contingent factors, in the end "the truth will out"—that is, in the words of Gilbert and Mulkay (1984: 94), "Social factors, personal judgements, intuition, charisma and so on are all allowed to play a part in science. But only in the short run. In the long run, it is scientific *facts* which are important." However, popular science narratives of discovery do not always focus on facts alone. By using presented discourse to showcase the contingent factors, they sometimes draw the focus away from the discoveries to the discoverers. Even though dialogue included in the narratives of discovery tends to show scientists engaged in discussions related to scientific activities, the narratives (by their nature) remain people-centered. So while they do not necessarily distort the information, they present it in a subjective way. I suggest this is a result of fictionalization through presented discourse.

As I have pointed out in the section discussing fictionality, fictionality does not have to stand in opposition to science and scientific inquiry. As Barberousse and Ludwig (2009) and Toon (2012) suggest, fictionality is an inevitable part of doing science. Gilbert and Mulkay (1984) and Harré (1994) show that there is a place for emotionality in professional science. Suarez (2009: 6) acknowledges that imagination is also an important aspect. Rouse (2009: 52) and Barberousse and Ludwig (2009: 57) point out that fictionality in professional science aids in establishing relations between phenomena and allows scientists to present new knowledge more easily.

The function of knowledge presentation is essential to popular science. By fictionalizing narratives of discovery, the authors help their readers process abstract scientific issues since the stories now are not solely about science but also about scientists. The face of professional science is no longer emotionless; science and fiction cross paths. These connections become evident in the analyses of fictionality in professional and popular science.

Conclusion

This study looks at the image of scientists as it is conveyed to the public through popular scientific writing. In the course of this book, I examine presented discourse of scientists in the narratives of discovery and discuss it in the light of fictionality. I approach fictionality as not necessarily dealing with imaginary entities and spaces or with hypothetical scenarios exclusively but as an expression of dramatization and emotionality that helps create subjective personal experiences associated with real-life events. These experiences emerge through presented discourse of scientists.

It may appear counterintuitive to use fictionality and emotionality in an analysis of essentially scientific subject matter and writing. However, both have a place in professional science, and popular science draws heavily on these less often discussed aspects of doing scientific research. I argue that presented discourse of scientists is used to re-introduce emotionality and subsequently fictionality into the narratives of discovery, and by doing so presented voices make the narratives resemble literary fiction while retaining scientific subject matter. It also becomes clear that popular science functions as one of the publication outlets of the scientific community, and as such, popular science paints a picture of professional researchers in an almost exclusively positive light, placing a greater focus on the successful results than on the challenges of the process by which these results are achieved.

The investigation of presented voices introduced here is based on the models of presented discourse analysis created by Leech and Short (1981) and by Semino and Short (2004). And while my analytical model (which appears in the following chapter) owes to the above mentioned researchers, it is an independent framework designed to account for the phenomena found in popular scientific writing and sometimes questioning the applicability of the traditional categories. That being said, the overview of the frameworks and the categories of presented discourse supplied in this chapter is a useful resource since I do not introduce new terminology.

CHAPTER 3

Analytical Background, Framework, Texts, and Methods

Introduction

This chapter discusses the technical aspects of presented discourse analysis in popular science. It introduces my own analytical framework and compares my findings to those of Semino and Short (2004). The goal is to place my data within the existing knowledge base on the subject. Since Semino and Short (2004) explore presented discourse in both fiction and non-fiction, it is important to understand how presented voices of scientists are expected to function in each type of text before any comparisons of the data take place.

Fu and Hyland (2014: 127, 141) assert that authors of popular science rely heavily on outside voices to make their arguments. Popular science, Fu and Hyland (2014: 141) argue, delivers its message not so much through narration but through the voices of the scientific community. As Fu and Hyland (2014: 141) declare, "scientists become real actors" in popular science stories of discoveries. The reliance on presented discourse is not surprising. However, its role in popular science is different from what has been suggested (see, for example, Fu and Hyland 2014, de Oliveira and Pagano 2006).

The idea that popular science, for the most part, borrows the rhetorical resources available to professional scientific writing and "shapes these for a lay audience" (Fu and Hyland 2014) sounds plausible especially in light of the argument introduced in the previous chapter. On the one hand, if popular science is a publishing outlet of the scientific community, it is expected that some of the rhetorical tools remain the same. After all, as Bucchi (1998: 12) suggests, popular science can be used for professional communication of the research results. In that case, the familiarity of the rhetoric is an advantage.

On the other hand, popular science has a more pronounced goal of promoting the positive image of the scientific community than professional science does. Therefore, it is understandable that the writers come up with new ways to implement the familiar rhetorical strategies. Presented discourse of scientists is one of these rhetorical tools that takes on new functions. Presented discourse has a markedly different role in popular science from the one it performs in professional writing. Fictionality introduced in presented discourse makes it resemble presented discourse in fiction.

© KONINKLIJKE BRILL NV, LEIDEN, 2018 | DOI: 10.1163/9789004365971_004

General Role of Discourse Presentation in Fiction and Non-fiction

The differences between fiction and non-fiction might seem obvious, and the language used is inevitably judged as one of the differences. At the same time, it is not entirely true that the language of fiction is radically different from that of non-fiction. When it comes to science writing in particular, the borders can be less rigid. Science writing comes in a variety of forms, both fiction and non-fiction. While the non-fiction forms—research articles, monographs, textbooks, popularizations—are more well-known, the fiction outlets devoted to science—novels, plays, and poetry describing the world of the scientific community—have received less attention. Both, however, share certain linguistic features, and in some cases, the fictional accounts (known under the genre term lab lit) offer more realistic descriptions of laboratory life (see, for example, Pilkington 2017a and Pilkington, O. and Pilkington, A. forthcoming). Separating fiction from non-fiction purely on linguistic grounds, therefore, is not as productive as it might appear (Lamarque 2000, Skov Nielsen et al. 2015a).

However, it is hard to deny that the same linguistic elements can be used differently in fiction and non-fiction. Presented discourse demonstrates a clear example.

Firstly, discourse presentation in fiction aids in the creation of experientiality, while in non-fiction presented discourse serves to boost accuracy and truthfulness. As Toolan (2001: 129) notes, presentation of speech in literary texts contributes to the authenticity of the story world, and Herman (2009: 147) points out that in fiction, "a rich context of felt experience emerges" as a result of "character's conversation." Semino (2004: 436–437) regards character voices as vital to the development of an emotional attachment between the reader and the characters. In fiction, presented discourse can easily project consciousness and help the reader see the events through the eyes of the characters, evoking empathy and sympathy—vital features of emotional engagement according to Toolan (2011). Discourse presentation contributes to what Fludernik (1996) identifies as a key feature of fictionality for literary texts—evocation of human experience.

In non-fiction (that is writing that excludes novels, short stories, and other works that are predominately imaginative), including discourse of others serves a purpose different from that of presented discourse in fiction. Semino and Short (2004: 226), having analyzed a corpus of fiction and non-fiction twentieth-century texts, conclude that fiction relies more on those properties of discourse presentation that emphasize dramatization of the events and inner worlds of the characters. In contrast, non-fiction makes greater use of information-carrying properties of presented discourse such as the ability to summarize.

There are multiple reasons for this preference. For instance, the voices which are introduced into an academic text are entirely subject to the needs of the author and do not contribute to the creation or representation of the identities of those who originated them (Livnat 2012). References to, and presentation of, multiple voices in academic writing fulfill the following functions according to Livnat (2012: 64–66): firstly, it is the establishing of a research context; secondly, an acknowledgement of the connection with the existing claims/knowledge; and thirdly, the construction of the author's research identity. In other words, presented discourse could be used as a background for the author's ideas.

What is a valued feature of presented discourse in fiction—invitation to the reader to engage with and to interpret the presented voices—is to be approached with caution in non-fiction. As Livnat (2012: 59) notes, "In scientific writing, the act of handling other speaker's utterances is less free than in other genres." As a result, when a new voice is introduced, the author, according to Livnat (2012: 59) is obligated to include his/her own interpretation which is to be adhered to by the reader as well. As de Oliveira and Pagano (2006: 641) note, the interpretation of presented discourse supplied by the author contributes to the dialogue between the author and the reader of non-fiction. Presented discourse in non-fiction is ultimately author-centered. It is not used to create characters or to establish a connection between the people whom presented discourse represents and the reader.

The focus on the author as the interpreter of presented voices can be found outside the professional scientific publications, in newspaper reports, for example (see Calsamiglia and Ferrero 2003, de Oliveira and Pagano 2006, Smirnova 2009). Presented discourse in such texts is seen as a vehicle for the author's opinions and interpretations not primarily as the means of characterization; it is used to facilitate a dialogue between the author and the reader.

Sometimes, this dialogue is directed at explaining complex information to the reader. Quite often, non-fiction (in an attempt to explain the world) has to introduce multiple voices in such a way that their messages are coherent to a wide audience who might not be well-versed in a particular issue. This is the case with science popularization. In this situation relying on the information-carrying and summarizing properties of discourse presentation accomplishes what Ciapuscio (2003: 210) calls "recontextualizing and reformulating one's source in such a way that it is comprehensible and relevant for a different kind of addressee." It becomes clear that in non-fiction, discourse presentation is used primarily to convey factual information. This function, it appears, dictates the form presented discourse is likely to take. For example, as Semino and Short (2004) have found, presented discourse in non-fiction is more likely to be indirect.

To summarize briefly, presented discourse in fiction is directed towards an expression of emotion and dramatization of the described events. Its main goal is to create an emotional response in the reader. Presented discourse in non-fiction is used to boost the accuracy of the account and is directed to the incorporation of facts. In light of this, presented discourse in fiction becomes the means for character creation (see, for example, Toolan 2001), while in non-fiction it is more likely to be the means of constructing the author's professional identity (see, for example, Livnat 2012).

Popular science combines the function of presented discourse in fiction with the preference for the form most often associated with non-fiction. That is, popular science narratives of dicovery rely more heavily on indirect types of presented discourse but use their summarizing properties to create dialogue and portray scientists as emotionally relatable characters.

The Framework for Presented Discourse Analysis of the Popular Science Corpus

The study described here and the subsequent discussion of presented discourse in popular science books focuses on the functions the voices and thoughts of scientists perform in such texts. That is not to say that I am not concerned with the forms presented discourse takes but to indicate that it is not my primary goal to introduce any new categories. Rather, the approach I take examines the interaction and the effects of the already existing categories as they were defined and discussed in chapter 2 (See Presented Discourse: An Overview of Analytical Approaches). With that in mind, I propose a simplified analytical model that is more suitable for the needs of the study.

I should also mention that I will follow the nomenclature suggested by Short (2012) and use the term "Presentation" [P] rather than "Representation" [R] in the names of the categories. The discourse presentation categories that I use are introduced below in tables 3.1.a, 3.1.b, and 3.1.c.

The arrangement indicates a progression from the categories where the author has the most control (Indirect Discourse, with NPSA, NPTA, NPWA being the most author-controlled categories) to the categories where authorial presence is minimal (Direct Discourse). The parentheses in the (F)DS category indicate that no difference between FDS and DS is observed. The same is true for thought presentation and writing presentation.

I should note that for the main portion of the analysis, I combine the categories of speech presentation and writing presentation. I will use the label

TABLE 3.1.A *Presentation of speech in popular science corpus*

Speech presentation scale			
Narrator's Presentation of Speech Acts (NPSA)	Indirect Speech (IS)	Free Indirect Speech (FIS)	(Free)Direct Speech [(F)DS]

TABLE 3.1.B *Presentation of thought in popular science corpus*

Thought presentation scale			
Narrator's Presentation of Thought Acts (NPTA)	Indirect Thought (IT)	Free Indirect Thought (FIT)	(Free)Direct Thought [(F)DT]

TABLE 3.1.C *Presentation of writing in popular writing corpus*

Writing presentation scale			
Narrator's Presentation of Writing Acts (NPWA)	Indirect Writing (IW)	Free Indirect Writing (FIW)	(Free)Direct Writing [(F)DW]

"speech" in the category names and the general term *presentation of communicated discourse* to indicate the combined type of spoken and written discourse.

The idea behind the creation of presentation of writing as a separate type of discourse presentation was accuracy—Semino and Short (2004: 48) note, for example, that Direct Writing may produce a more "accurate word-by-word representation" than Direct Speech. However, Semino and Short (2004: 50) point out that "The writing presentation scale is very like the speech presentation scale in relation to the effects associated with particular categories. This is … because in both cases the original is … a piece of discourse, even though the medium is different." Short (2007: 230 and 2012) endorses this view of the two categories.

While I see the potential usefulness of the distinction (and return to it in chapter 6), in my popular science corpus of narratives of discovery, it is sometimes impossible to distinguish between these two categories, as the authors may present writing in the form of speech by using the scientists' diaries, papers, or even secondary publications as sources for presented discourse, not to mention that often no specific source for a particular instance of discourse is identified. For example, Greene's (2011: 11–12) narrative of Lemaître's discovery of the Big Bang, introduced here as example 3.1, includes several instances of speech presentation from Lemaître and Einstein. The fragments in blue identify the utterances as speech since they introduce locations and situations where people are more likely to talk to each other than to exchange written messages, and in the case of the last fragment in blue, the combination of the physical action and the reporting verb suggests speech rather than writing:

3.1 "Your mathematics is correct, but your physics is abominable." The 1927 Solvay Conference on Physics was in full swing, and this was Albert Einstein's reaction when the Belgian Georges Lemaître informed him that the equations of general relativity ... entailed a dramatic rewriting of the story of creation ... The universe, Einstein admonished Lemaître, is not now expanding and never was ... Six years later, in a seminar room at Mount Wilson Observatory in California, Einstein focused intently as Lemaître laid out a more detailed version of his theory ... When the seminar concluded, Einstein stood up and declared Lemaître's theory to be "the most beautiful and satisfactory explanation of creation to which I have ever listened."

Note that all the indications of speech presentation (rather than writing) come from the author. They either set the scene (conference, a seminar room) or are reporting devices ("stood up and declared"). Presented discourse itself contains no indications of being either speech or writing. Only one instance of Direct Speech (green) is attributed to a source (a secondary one) in the chapter end notes. The reference is insufficient in determining whether the original words were expressed orally or in writing. The shift from the third to the first person narration that the instance of Direct Speech creates changes the perspective of the story but does not help identify this particular stretch of presented discourse as either speech or writing. Direct Speech in example 3.1 can be easily compared to Direct Speech in a novel, where the "idea of anterior vs. posterior discourse situations does not sensibly apply at all" (Short 2012: 20) since it is impossible to determine what the original was. In my opinion, the distinction between speech and writing in this case is not important since as Semino and Short (2004) and Short (2012) note, the effects each produces are similar.

The same situation occurs in Kean (2012: 138–141) in his narrative of Rouse discovering that a virus can cause cancer. Speech presentation appears without any references in the chapter, but the end notes indicate that a biography of the scientist was used in creating the story; the reference is not specific enough to identify which instances of the speech presentation it covers, and it is certainly not enough to distinguish between writing and speaking. This makes it problematic to check the potential instances of writing presentation for accuracy—the chief parameter for the division between writing and speech (see Short et al. 2002: 327, Semino and Short 2004: 113).

Overall, the majority of the reporting verbs in the narratives of discovery are non-specific enough to allow presented discourse to be interpreted as either speech or writing. The very presence of the ambiguity suggests the lack of importance of the distinction in the narratives. For these reasons, I will use the label *presentation of communicated discourse* (or *communicated discourse*) to cover both speech and writing presentation. In distinguishing between speech and writing on the one hand and thought on the other, I am in accord with Short (2012: 22), who notes that speech and writing have communicative functions, while thought is reflective of a private inner world. Short (2012) also combines the categories of speech and writing presentation for the purposes of his analysis; however, he does not give a specific name to the combined category and refers to it as "speech and writing."

Another adjustment I make is the merging of Semino and Short's categories of Narrator's Representation of Voice and Narrator's Representation of Thought (the categories prototypically associated with production of minimal summaries of discourse) with Narrator's Presentation of Speech Acts and Narrator's Presentation of Thought Acts respectively. They differ by the degree of summarizing they produce. While degrees of summarizing could be important, for my purposes, a general indication of a summary is sufficient. When discussing summaries, I will rather focus on whether they are created by means of direct or indirect discourse than on the degree of summarizing. Thus I will consider both of the following examples NPSA even though one introduces more details than the other:

3.2.a Conway related the story of searching for the Higgs boson in Fermilab data (the LHC wasn't online yet) using his personal favorite channels, ones where a tau lepton is produced.

CARROLL 2012: 199

3.2.b He said no.

KAKU 2011: 130

Both 3.2.a and 3.2.b are summaries created using indirect discourse. What is important to me are the summarizing properties of both instances of presented discourse not so much the degree of detail or to what extent the author had condensed the original.

The categories introduced in the Semino and Short (2004) multi-genre corpus study offer some very fine distinctions, which are less relevant in analysis of a more homogenous group of texts. Previous analyses of speech and thought presentation in a single genre (see, for example, Waugh 1995, Myers 1999, Toolan 2001, Urbanova 2012) tended to have a narrower focus. Waugh (1995) analyzes only Direct and Indirect Speech in newspaper reports. Myers (1999) chooses to emphasize direct discourse only when examining presented speech in oral group discussions. Toolan (2001) suggests that it is possible to omit Internal Narration and Narrator's Representation of Voice from the discourse presentation scales when analyzing fiction; he also draws attention to the more broad distinction between direct and indirect discourse—something that Semino and Short (2004) do not explicitly emphasize. Urbanova (2012), looking at newspapers, sees it sufficient to examine only free direct and direct forms of presented discourse, which also suggests an underlying broad contrast between direct and indirect discourse.

The one category introduced by Semino and Short (2004) that I found more relevant than other additions is Narrator's Report of Speech/Narrator's Report of Thought (NRS/NRT). I did not include it on the scales because it represents not so much presented discourse itself but the reporting signals used to introduce it. Short (2007, 2012), for instance, also omits it from the scale. This category accounts for a broader range of phenomena than traditional reporting clauses. Of all the new categories proposed by Semino and Short (2004), it is the only one that broadens the reach of an older counterpart—reporting clause.

In general, I find the categories with a broader reach more useful for my analysis, as I focus on the interaction and functions of discourse presentation types. Semino and Short (2004) and Short (2007, 2012) identified the issue of interaction among the types of presented discourse labeling the instances "embedding," "quotation phenomenon" (Semino and Short 2004), "discourse-domain summary" and "quotative summary" (Short 2012). However, they did not analyze these interactions and combinations of presented discourse types for their dramatization properties. (I explore this issue in chapter 4.) The same is true for the issue of fictionality. Even though Semino and Short (2004) acknowledge the presence of dramatization, emotionality, and personal perspective in presented discourse, they do not discuss these as manifestations of fictionality in non-fiction. At the same time, the conclusion that Semino and Short (2004: 170–171) draw from the analysis of presented discourse is important for the point this study makes—the "differences in SW&TP [Speech/Writing & Thought Presentation] ... do not necessarily contrast the fiction section with the two non-fictional sections, but suggest much more complex similarities and contrasts among different text-types" (Semino and Short 2004: 170). Semino and Short do not venture beyond this acknowledgement, but I hope that my analysis of fictionality illuminates these aspects of presented discourse in non-fiction.

Corpus Selection and Methodology

Why Study Popular Science Narratives?

Narrative is a popular form of knowledge dissemination. In fact, some scholars (see, for example, Schank 1990; Gjedde 2000; Herman 2009) suggest that it is the preferred method for humans to acquire and process new information. Herman (2009: 9) writes that "... human intelligence itself is rooted in narrative ways of knowing, interacting, and communicating." In other words, human knowledge is structured, stored, and shares in a form of narrative.

Another view on the power of narrative suggests that the value of a story is not necessarily in its construction and storage of knowledge but in a narrative's ability to supply context necessary to interpret facts (see, for example, Turney 2004b, Reitsma 2010; Avraamidou and Osborne 2008). This view of narrative usually predominates in discussions of popular science texts. As Reitsma (2005: 93) notes, "A narrative ... includes interpreted information," arguing that this eliminates the need to draw conclusions between the events presented. Turney (2004b: 333) surveys the kinds of narratives employed in popular science and suggests that they all serve one purpose—to explain or "translate" science into laymen's terms. This interpretation of the role popular science narratives play can be described as the "scaffolding view."

On the other hand, Eggins and Slade's (1997) research into casual conversation suggests another function of narratives. Eggins and Slade (1997: 229) demonstrate that narratives reveal "values, attitudes, and ways of seeing the world." This approach can be described as the "ideological view" and applies not only to spoken, dialogic narratives (as Eggins and Slade show) but also to the written, monologic narratives, such as popular science narratives of discovery.

In fact, the ideological function of the narratives of discovery is quite pronounced, as these stories are used not only to explore and explain scientific concepts but also to showcase scientists as capable and creative individuals, immune to failure. As the following chapters show, presented discourse of scientists is used to deliver not only the scientific message but also to project emotion, create connections, and to characterize the members of the scientific community as relatable and interesting individuals. In other words, the scientists in the narratives as seen through the lens of discourse presentation closely resemble characters out of literary stories rather than appear as disembodied voices of reason delivering cold facts to the reader.

The form in which an account of a discovery is delivered can influence the readers' view of science. A story is designed to focus on the people in it and as such is an ideal vehicle for the celebration of the human side of science. The narrative form, however, does not automatically exclude objectivity, but it does lend itself to the promotion of a positive message more so than a differently structured account would (for more on that see Appendix).

Description of the Popular Science Corpus

The study presented in this book is based on an analysis of 100 narratives of discovery found in popular science books. 50 narratives were extracted from books written by scientists and 50 from books written by science journalists. I initially separated the corpus into the two subgroups anticipating differences in the numbers, types, and treatment of presented discourse. This hypothesis was based on de Oliveira and Pagano's (2006: 642, 643) findings that show discrepancies in the treatment of presented discourse by professional scientists and science journalists. However, upon closer analysis, my corpus did not supply the evidence to support their observations. The reason might be the difference in the texts examined. While de Olivera and Pagano (2006) looked at articles—professional research articles and popular scientific articles—my corpus is based on popular science books. Therefore, the final version of my analysis does not distinguish discourse presentation based on the qualifications of the author.

10 books were used to construct the corpus. They are listed below alphabetized by the author's last name:

- Bill Bryson 2003, *A Short History of Nearly Everything*
- William Bynum 2012, *A Little History of Science*
- Sean Carrol 2012, *The Particle At the End of the Universe*
- Enrico Coen 2012, *Cells to Civilization*
- Marcus du Sautoy 2011, *The Number Mysteries*
- Timothy Ferris 1988, *Coming of Age in the Milky Way*
- Brian Greene 2011, *The Hidden Reality*
- Michio Kaku 2011, *Physics of the Future*
- Sam Kean 2012, *The Violinist's Thumb*
- Carl Zimmer 2011, *A Planet of Viruses*

These texts represent the following scientific disciplines: astronomy, chemistry, genetics, mathematics, medicine, physics, and virology. All of them are originally written in English, and most have been published between 2011 and 2012. However, the study also incorporates two older texts: one from 2003 and one from 1988. The reason behind including older texts was to see whether examples taken from them fit within the analytical categories adopted primarily with the newer texts in mind. For example, one of the underlying claims that I make is that popular science is directed towards the presentation of scientific issues through emotionally engaging mechanisms, especially dramatizing. Using older texts allows me to say that such a mode of presentation is not as novel as current research of popular science suggests. It was probably underreported by the analysts, who, as my overview of the field leads me to believe, became focused on the idea of the emotional connection with science fairly recently.

Overall, even though I do supply the frequency counts for the types of the discourse presentation observed, this is largely a qualitative study with a sample of 193 occurrences of the discourse presentation of scientists. These occurrences are short stretches of discourse that are on average about 35 words long. However, the narratives themselves are not very long either, ranging from between 200 to 500 words. The narratives of discovery are not themselves the end products but rather brief excursions into history on a larger journey to uncovering the potentials of modern-day science. They often provide the reader with the much-needed background on fundamental laws and scientific principles.

In order to collect the sample of the discourse presentation occurrences that was representative of each author but at the same time still manageable for manual analysis, I first limited the number of the narratives analyzed from each text. I looked at 10 narratives of discovery from each author. To collect the 10 narratives from each text, I first went through each book and identified all the narratives of discovery, using the specific criteria. Narratives of discovery possess a certain structural makeup: they start out with a hypothesis, develop/prove/disprove it, and then supply an evaluation, which examines the discovery's relevance. Such narratives often function as supplements for the scientific concepts discussed in the books.

Such identification is based on using Labov's (1972) structural model to separate popular science narratives of discovery from other portions of popular science books. When applied to certain segments of popular science texts, Labov's model not only supports the claims for narrativity but also determines that the structural makeup of the narratives is dependant on the explanatory function of these text segments (see Pilkington 2017b and Appendix).

From the books that included 15 or fewer narratives of discovery, the first 10 narratives were taken. From the books that included more than 15 narratives of discovery, I used three narratives from the beginning of the text, three from the middle, and four from the concluding sections.

Having selected the narratives, I then analyzed each narrative for evidence of presented discourse of scientists. After that I proceeded to identify all occurrences of communicated discourse and thought presentation, labeling them according to the categories introduced in the previous section. In addition to the discourse presentation categories, I also analyzed the Narrator's Reports of Speech/Narrator's Reports of Thought (reporting clauses) for the presence of emotionality that projects the authors' feelings toward presented discourse that follows.

NRS/NRT have been suggested (see, for example, de Oliveira and Pagano 2006: 641) as mechanisms that place the "writers and readers in interactive relation" by allowing the authors to evaluate the stretch of presented discourse that follows. Thus authors can guide the readers' perception of and emotional reaction to presented discourse through specially marked NRS/NRT. Consider example 3.3 (NRS is in blue):

3.3 Then, in 2010, scientists made the shocking announcement that 5 percent of the debris contained water, so the moon was actually wetter than parts of the Sahara desert.

KAKU 2011: 268

By evaluating the discovery as "shocking" before presenting the actual details of the announcement, Kaku is influencing the readers' emotional reaction to the discovery. The readers are to perceive the findings as unexpected and unusual. In cases like these, especially where presented discourse is indirect, the author takes it upon himself to convey the emotionality of the message possibly because the original utterance did not possess the same level of emotion. This could be also seen as the author's monitoring of the understanding and evaluation of the discovery by the reader (The notion of monitoring is usually associated with fiction; see, for example, Murphy 2005.). I shall refer to such emotionally marked NRS/NRT as NRS_e/NRT_e, where "e" indicates emotionality (see Chapter 4).

My analysis yielded 193 total occurrences of speech and thought presentation (not counting NRS/NRT as separate occurrences) in 100 narratives of discovery. While the number of occurrences is not very large, the sample is sufficient to demonstrate that popular science relies on dramatization and fictionality in order to produce an almost entirely positive image of the scientific community.

Comparison of Frequency Information for the Popular Science Corpus with the Semino and Short Findings

The Semino and Short (2004) study is the most comprehensive corpus analysis of presented discourse in non-fiction to date even though it does not deal with non-fiction exclusively. The size of the corpus (258,348 words and 16,533 occurrences of presented discourse) allows for their findings and conclusions to be representative of the genres they examined. At the same time, I remain mindful that Semino and Short (2004) included only certain genres to represent fiction and non-fiction. For instance, the fiction section was represented by novels and the non-fiction section by newspapers, biographies, and autobiographies. While it is possible that the newspaper reports analyzed contained scientific news and thus represented the genre of popular science, there is no clear indication of that in Semino and Short (2004). Therefore it is unclear if their findings can be generalized to include the genres beyond those examined. In investigating popular science, another non-fiction genre, I am testing their observations and conclude that for the most part their findings are more generalizable than Semino and Short (2004) themselves were able to argue based on the limitations of their corpus.

In this study, I suggest a combined category presentation of communicated discourse to cover both speech and writing presentation and to contrast with presentation of thought. I will continue to use the terms "communicated discourse" and "thought presentation" throughout the book when referring to my own findings, but I will use "speech/writing presentation" when discussing research of others who employed these specific terms. I will also continue to use the abbreviation "R" (Representation) instead of "P" (Presentation) in names like Narrator's Representation of Speech Acts (NRSA), for example, when discussing the findings of Semino and Short (2004) since this was the term they used. However, when talking about my own findings, I will use the more recent term "Presentation" and the abbreviation "P." The names of the categories in the tables and the table titles will also reflect this discrepancy. In the frequency information introduced below, the numbers for the non-fiction section of the Semino and Short (2004) corpus are a combination of the figures for their press and (auto)biography sections unless otherwise identified.

Compared to the vast Semino and Short (2004) corpus, my sample is rather small. I analyzed a total of approximately 30,000 words and 193 occurrences of presented discourse. At the same time, my smaller corpus is much more specialized, dealing not only with just one non-fiction genre but also focusing on a very specific subset of texts within it—narratives of discovery. Nev-

TABLE 3.2.A *Frequency of communicated discourse presentation and thought presentation in the narratives of discovery*

Total number of discourse presentation occurrences	
193	
Communicated discourse # of occurrences/ percentage	**Thought presentation # of occurrences/percentage**
140/72%	53/28%

TABLE 3.2.B *Semino and Short's frequency counts for speech/writing and thought presentation*

	Speech/writing	Thought
Whole corpus	58.62%	16.24%
Non-fiction [press]	70.65%	5.66%
Non-fiction [(auto)biography]	53.89%	18.23%
Fiction	51.71%	24.40%

SOURCE: SEMINO AND SHORT 2004: 59

ertheless, my data correlates with the general findings about the distribution of presented discourse observed by Semino and Short (2004). For example, Semino and Short (2004: 59) demonstrate that the speech/writing presentation occurs more often than the presentation of thought. Tables 3.2.a and 3.2.b show the correlation between my data and that of Semino and Short (2004).

I should note that Semino and Short (2004) included narration and what they called "Portmanteau" tags as part of their frequency counts. Portmanteau tags were assigned to stretches of presented discourse that contained "ambiguities within each mode of presentation (e.g. DS-FDS)" (Semino and Short 2004: 58). I did not include narration as part of the frequency counts and did not consider portmanteau tags. This is why my percentages of communicated discourse (speech/writing) and thought presentation frequencies add up to one hundred, while Semino and Short's (2004) percentages for speech/writing and thought presentation do not. Taking these discrepancies into account, the over-

TABLE 3.3 *Speech/writing to thought presentation ratios (Semino and Short 2004) compared with communicated discourse to thought presentation ratios*

Ratio of Speech/writing to Thought Presentation in Semino and Short's (2004) Fiction Section	2.1 : 1
Ratio of Communicated Discourse to Thought Presentation in the Narratives of Discovery	2.5 : 1
Ratio of Speech/writing to Thought Presentation in Semino and Short's (2004) (auto)biography Section	3.0 : 1
Ratio of Speech/writing to Thought Presentation in Semino and Short's (2004) Whole Corpus	3.6 : 1
Ratio of Speech/writing to Thought Presentation in Semino and Short's (2004) Press Section	12.5 : 1

all picture is the same for both studies: thought presentation is significantly less frequent. These numbers also show that the relationship between thought presentation and presentation of communicated discourse in the narratives of discovery is somewhere between that of speech/writing and thought in the (auto)biography section and in the fiction section of the Semino and Short (2004) corpus. The ratio of communicated discourse to thought presentation in the narratives of discovery is 2.5. The ratio of speech/writing to thought presentation in the (auto)biography section of the Semino and Short (2004) corpus is 3, and in the fiction section it is 2.1. See table 3.3.

Presentation of Communicated Discourse in the Narratives of Discovery and Speech/Writing Presentation in Semino and Short (2004)

Despite the great discrepancies in the number of occurrences between my corpus and the Semino and Short (2004) data, the general tendencies of discourse presentation for non-fiction are the same, as percentage points in table 3.4 (see below) clearly demonstrate.

As Table 3.4 shows, the percentages for the frequency of the specific discourse presentation categories in my corpus and in that of Semino and Short (2004) are rather similar. In fact, if my data is compared to the Semino and Short (2004) non-fiction section (see table 3.5 below), the percentages are almost identical: with (Free)Direct Speech and Narrator's Presentation of Speech Acts (NV+ NRSA) dominating the Public Discourse presentation spectrum. In both the Semino and Short (2004) study and in my corpus, the most and the least author-controlled types of presented discourse, (F)DS and NPSA,

TABLE 3.4 *Comparison between presentation of communicated discourse frequencies in the narratives of discovery and the speech/writing frequencies in the Semino and Short (2004) whole corpus*

Type	Narratives of discovery # of occurrences followed by percentage	Semino and Short (2004) # of occurrences followed by percentage
(F)DS/(F)DW	47 (33%)	3155 (47%)
IS/IW	34 (25%)	1188 (18%)
FIS/FIW	4 (3%)	188 (2.8%)
NPSA (NV/NW + NRSA/NRWA)	47 (33%)	2045 (31%)

SOURCE: SEMINO AND SHORT 2004: 67, 101

TABLE 3.5 *Presentation of communicated discourse scale with percentages from Semino and Short (2004) non-fiction section [1] and from the narratives of discovery [2]*

NPSA—IS—FIS—(F)DS
[1]38%—24%—3%—35%
[2]33%—25%—3%—33%

account for a little over 1/3 of the total occurrences each. This means that more than 60% of communicated discourse and speech/writing is located at the extremes of the discourse presentation scale.

These percentage points show that discourse presentation in the narratives of discovery corresponds closely to the Semino and Short (2004) non-fiction data. Such correlation was expected since the narratives of discovery fall within the genre of non-fiction.

Presentation of Thought in the Narratives of Discovery and in Semino and Short (2004)

Unlike presentation of communicated discourse, presentation of thought in the narratives of discovery does not follow the overall pattern established by Semino and Short (2004) very closely. See Table 3.6. However, as the data in

TABLE 3.6 *Comparison of thought presentation frequencies in the narratives of discovery with thought presentation frequencies in the Semino and Short (2004) whole corpus*

Type	Narratives of discovery # of occurrences followed by percentage	Semino and Short (2004) # of occurrences followed by percentage
(F)DT	3 (5.6%)	107 (15%)
IT	24 (≈47%)	201 (29%)
FIT	0	275 (40%)
NPTA	26 (≈47%)	114 (16%)

SOURCE: SEMINO AND SHORT (2004: 115)

TABLE 3.7 *Comparison of thought presentation frequencies in the narratives of discovery with thought presentation frequencies in the Semino and Short (2004) non-fiction section*

Type	Narratives of discovery # of occurrences followed by percentage	Semino and Short (2004) # of occurrences followed by percentage
(F)DT	3 (5.6%)	30 (13%)
IT	24 (≈47%)	106 (46%)
FIT	0	45 (19%)
NPTA	26 (≈47%)	52 (22%)

SOURCE: SEMINO AND SHORT (2004: 115)

Table 3.7 demonstrates, the distribution of thought presentation in the narratives of discovery resembles the pattern for non-fiction observed by Semino and Short (2004).

Semino and Short's (2004) frequency information for thought presentation in non-fiction and in the whole corpus suggests two different categories as the norm for presentation of thought: Free Indirect Thought for the corpus and as a whole (and for fiction) and Indirect Thought as the norm for non-fiction. FIT, the norm for Semino and Short's (2004) whole corpus, is not represented at all in the narratives of discovery. This might appear as a striking difference between the two corpora. However, when my data is compared with the non-fiction section of the Semino and Short (2004) corpus, the difference is less pronounced, especially if the non-fiction section is analyzed not as a combination of press and (auto)biography but as its component parts. According to Semino and Short (2004: 115), there are no occurrences of FIT in the press section. The 45 occurrences reflected in Table 3.7 are all from (auto)biographies. This helps reconcile my data with the figures from Semino and Short (2004) since, as they show, not every non-fiction genre includes FIT. However, FIT in the Semino and Short (2004) non-fiction section is not the least frequent category. In the whole corpus and in the non-fiction section of Semino and Short (2004), it is (Free)Direct Thought that occupies this position. If FIT is excluded from the consideration in my corpus, then the frequency count for (F)DT falls in line with what Semino and Short (2004) observed.

Conclusion

As the comparison between the frequency counts for presentation of communicated discourse and presentation of thought with the Semino and Short (2004) corpus indicates, the narratives of discovery distribute presented discourse across the available categories in a manner consistent with other non-fiction genres. That means that indirect forms of discourse presentation dominate. The only instance where this is not entirely so is in presentation of communicated discourse, where (Free)Direct Speech is found to be as frequent as Narrator's Presentation of Speech Acts. The difference between (F)DS and NRSA in the Semino and Short (2004) non-fiction section is 3%. While this is not a significant gap, its very presence allows Semino and Short (2004: 226) to conclude that "non-fictional genres make greater use of the less direct categories of speech presentation." At the same time, the data from the communicated discourse of the narratives of discovery does not lead to the same conclusion and points in favor of my initial assumption that popular science uses indirect discourse for the purposes usually assigned to direct discourse in fiction. I explore this assumption in the following chapter.

CHAPTER 4

Dramatization in the Narratives of Discovery: The Roles of Communicated Discourse and Thought Presentation

Introduction

The comparative analysis that concluded the previous chapter identifies notable points about the roles of presentation of communicated discourse and presentation of thought in popular science narratives of discovery. For instance, the frequency counts indicate that (Free)Direct Speech and Narrator's Presentation of Speech Acts are prominent in presentation of communicated discourse and that Indirect Thought and Narrator's Presentation of Thought Acts are important to presentation of thought. In this and the following chapter, I will analyze the forms and functions of communicated discourse and thought presentation and will examine not only the individual types but also address the issue of interaction of the various presented discourse types.

The present chapter focuses on dramatization achieved through the use of communicated discourse and thought presentation. To recap the definition introduced in Chapter 2, by dramatization I understand those aspects of fictionality that contribute to the re-imagining and reconstructing of real-life events and also those mechanisms that help show to the reader the subjective experiences of the scientists. From the point of view of presented discourse, re-imagined and reconstructed real-life events manifest through presentation of dialogue. Subjective experiences of the scientists are incorporated using emotionality and personal perspective through presented discourse.

When the categories of presentation of communicated discourse and presentation of thought are considered individually, it appears that only communicated discourse contributes to dramatization. For instance, a little over half of all occurrences of communicated discourse perform dramatizing functions, while no occurrences of thought presentation are used to dramatize. This breakdown is somewhat unusual since other studies suggest that presentation of thought has dramatizing properties, especially those associated with emotionality (see, for example, Semino and Short 2004: 118, 124; Short 2007: 231). On the other hand, thought presentation categories that are traditionally regarded in connection with "vividness" and "drama" (Short 2007: 231) or with "heightened emotion" and "protracted access to the consciousness of charac-

© KONINKLIJKE BRILL NV, LEIDEN, 2018 DOI: 10.1163/9789004365971_005

ters" (Semino and Short 2004: 118, 124) are practically absent from the narratives of discovery. These categories are Free Direct Thought, Direct Thought, and Free Indirect Thought (See Tables 3.6 and 3.7 in Chapter 3). There are only 2 occurrences of DT in the narratives, 1 occurrence of FDT, and 0 occurrences of FIT. At the same time, when presentation of thought is considered not separately but alongside presentation of communicated discourse, it does contribute to dramatization.

While I argue that all types of communicated discourse can potentially contribute to dramatization, my data suggests that (Free)Direct Speech and Narrator's Presentation of Speech Acts are the most prominent ways of fictionalizing the narratives of discovery. When it comes to presentation of thought, NPTA are most likely to combine with communicated discourse. There are 9 NPTA that work in combination with communicated discourse versus only 4 occurrences of IT that do so.

Taking into account that the majority of dramatization occurs through communicated discourse, the chapter will address presentation of communicated discourse in more detail, with NPSA and (F)DS being the focus. Other categories of communicated discourse as well as thought presentation will be mentioned in the discussion of the interaction among discourse types. However, they will receive more attention in the following chapter that addresses non-dramatizing properties of presented discourse in the narratives.

In the narratives of discovery the functions of NPSA are not limited to summarizing as has been suggested by previous analyses of speech and writing presentation (see, for example, Toolan 2001, Semino and Short 2004, Short 2007). Narrator's Presentation of Speech Acts can be used to dramatize as well as to summarize and is the preferred method of introducing dialogue in the narratives of discovery.

In effect, the data from popular science narratives of discovery suggests a re-evaluation of the function scale for presented speech/writing found in Short (2007: 230). Dealing with examples from literary fiction, Short (2007: 23) assigns dramatizing properties ("drama," "Showing") primarily to direct and free direct forms of speech/writing. He does, however, allow the possibility that Free Indirect Speech and Free Indirect Writing may also contain these but to a lesser degree. Thus, in a way, Short (2007) sets the precedent for indirect discourse possessing some level of dramatization. At the same time, FIS/FIW is the cut off point for the dramatizing properties on his functional scale—Indirect Speech and Narrator's Representation of Speech Acts are seen as fulfilling the summarizing functions only and thus distancing the reader from the characters and their actions. When it comes to popular science narratives of discovery, the dramatizing function is universal in presentation of communicated discourse and

TABLE 4.1.A *Functions of speech/writing presentation*

[N]	NV	NRSA	IS	FIS	DS (including FDS)
Summary					
Distance, calmness, etc.			Vividness, drama, etc.		
Telling					Showing

SOURCE: SHORT (2007: 230)

TABLE 4.1.B *Functions of communicated discourse presentation in the narratives of discovery*

NPSA	IS	FIS	(F)DS
Dramatization			
Summary		Vividness	

may be found throughout the scale but is concentrated mostly at the very ends of the scale with (F)DS and NPSA exhibiting more dramatizing properties than other types of communicated discourse.

This leads to a conclusion that the degree of dramatization is not tied to the directness or indirectness of a discourse presentation type but should be evaluated in each text or text type independently. Myers (1999: 397) comes to a similar realization regarding the functions of presented discourse. He points out, relying on his own evidence and on that of Baynham (1996) and Thompson (1996), that the functions of presented discourse "are more complex than was suggested by studies of … literary language, and that the functions vary with mode, genre, and social roles." Below I reproduce the function scale from Short (2007: 230)—table 4.1.a—and present my proposed scale that reflects the modifications suggested above—table 4.1.b.

In popular science narratives of discovery, NPSA retains their summarizing properties but may utilize them to dramatize. In the following sections, I supply examples and explain the reasoning behind the proposed changes to the function scale.

Dramatization in Presented Discourse: It is Not Just (F)DS

Semino and Short (2004: 96, 150–151) note that (F)DS is the most frequent category "by a long way" in the novels, biographies, autobiographies, and newspapers they examined.

(Free)Direct Speech and (Free)Direct Writing are used as often as all the other categories of presented discourse combined. In the fiction texts analyzed by Semino and Short, 73 % of all occurrences of presented discourse belonged to (F)DS. These numbers confirm Leech and Short's (1981/2007: 276) suggestion that Direct Speech is the norm for the presentation of speech.

On the other hand, Semino and Short's (2004) study and my analysis of the narratives of discovery highlight the prominence of another category of discourse presentation—Narrator's Presentation of Speech Acts (NPSA). In popular science narratives of discovery, NPSA are as frequent as (F)DS. In Semino and Short's (2004) study, the corresponding category—NRSA/NRWA—was the second most frequent after (F)DS (the tables in chapter 3 supply more precise data). The earlier study by Leech and Short (1981/2007: 260) that focused on fiction exclusively, designated this category of presented discourse as only "useful for summarizing relatively unimportant stretches of conversation." However, my data and that of Semino and Short (2004) shows a very high frequency Narrator's Presentation of Speech Acts, which suggests two possible conclusions: (1) either NPSA summarize more than just "unimportant" conversations or (2) they perform functions additional to summarizing. The data from popular science, in particular, [(F)DS and NPSA have the same number of occurrences, with each accounting for about 33 % of communicated discourse presentation] points to the importance of this category.

It is generally accepted that (F)DS is associated with dramatization (see, for example, Semino and Short 2004: 89, Short 2007: 230) and NPSA with summarizing (see, for example, Toolan 2001: 130, Semino and Short 2004: 73, Short 2007: 230). In popular science, both of these properties work together: the authors often draw on the summarizing properties of NPSA in order to dramatize.

Dramatization, as it relates to discourse presentation, is usually understood differently when it comes to fiction and non-fiction. However, it is most often associated with (F)DS no matter the text type. For example, in fiction (F)DS is used to create characters, achieve insight into their lives, and position the reader as an overhearer of character dialogue (Rimmon-Kenan 2002: 63–64; Semino and Short 2004: 92; Toolan 2001: 129). In non-fiction, dramatization, according to Semino and Short (2004: 94), is limited primarily to the presentation of personal perspective and the demonstration of emotions.

The dramatizing properties of (F)DS in non-fiction are secondary, eclipsed by the primary expectation of accuracy and faithfulness of representation (see, for example, Bell 1991, Semino and Short 2004, Urbanova 2012). The studies that suggest such breakdown of functions for (F)DS draw their conclusions primarily from the examples found in newspapers.

These findings on dramatization point to two possible outcomes for non-fiction: (1) there is very little dramatization in non-fiction, and (2) dramatization in non-fiction might be achieved through the means of discourse presentation other than (F)DS. It is important to keep in mind, however, that the narratives of discovery appear to differ from other non-fiction genres in the expectations of faithfulness for (F)DS. The absence of clear references to sources in the narratives suggests that dramatization is more important than reliability. In that, the narratives of discovery are closer to fiction than to non-fiction. Evidence suggests that there is no lack of dramatization in popular science but that it is divided between two discourse presentation categories—(F)DS and NPSA. The features of dramatization associated primarily with (F)DS in fiction—showing of characters' personalities/relationships—in the narratives of discovery manifest through NPSA. (F)DS in the narratives performs the functions usually associated with it in non-fiction—offering of personal perspective and addition of emotionality.

Dramatization through NPSA and NPTA

When (F)DS is associated with dramatization, it is mostly for its ability to create fictional dialogue. Example 4.1 shows a typical dramatization of dialogue in a novel. The particular example comes from Allegra Goodman's *Intuition*—a scientifically themed novel about researchers in a medical laboratory:

4.1 "This is not how we want to present ourselves."
"I know," said Cliff, "I didn't think—"
"Well, you should have thought about it."
"It's just a student paper," said Cliff. "And it's the summer weekly issue. Nobody's going to read it."
"Your interview is in the public record now," Glass snapped.
"You told me to meet with him!" Cliff burst out. "You asked me to speak with him."
"I assumed that while speaking to Jeff, you'd use your common sense."
"Look, he asked me about my role in the work. I just answered his questions."
"Your answers," said Glass, "do not match any of the other stories out there."

GOODMAN 2006: 167

There are no examples like this in my corpus, and the most typical use of (F)DS is to introduce single utterances as in example 4.2.

4.2 Rous himself later admitted, "I used to quake in the night for fear that I had made an error."

KEAN 2012: 140

This is precisely what Semino and Short (2004: 93) observed in the non-fiction section of their corpus—(F)DS "is normally used to present individual utterances in isolation." At the same time, the absence of dialogue conveyed through (Free)Direct Speech does not mean that there is no dialogue in the narratives of discovery. Dialogue found in popular science texts is, however, of a different nature—both formally and functionally. Formally, it is introduced using Narrator's Presentation of Speech Acts and not (F)DS, and functionally, it is often employed in order to dramatize not only the immediate events but also to re-imagine scientific debates and present them as dynamic verbal exchanges rather than as a series of publications or talks that took place over a lengthy period of time. In the process of these dramatizations, the authors project the personalities and the relationships of the scientists.

In fact, if dramatization is equated with presentation of dialogue [the prototypical function of (F)DS in fiction and thus the prototypical way of looking at dramatization], then it is possible to make a claim that in the narratives of discovery NPSA alone perform the dramatizing function since the majority of the dialogue, by far, is introduced via NPSA. There are 18 NPSA that in one way or another contribute to interactive exchanges, while (F)DS alone is never used to create dialogue, and there is only one instance of dialogue presented as IS. Those NPSA that do not present interactive exchanges function as prototypical NPSA that supply summaries. Most often (17 out of 29 non-dramatizing NPSA) they deal with scientific ideas and concepts proposed by scientists. However 12 out of 29 non-dramatizing NPSA showcase the scientists' reaction to the research of their colleagues or their own findings, thus still contributing to character creation. I discuss the non-dramatizing NPSA in the next chapter.

There are four subtypes of NPSA that are used to create interactive exchanges between the characters in the narratives. These are: (1) dialogic NPSA—one NPSA that incorporates two distinct voices in dialogue with each other, (2) strings of individual NPSA that together form a backbone of a dialogic exchange, (3) borderline dialogic NPSA—these either name a second interactant but do not give voice to him/her, or name the people involved in an interaction without introducing their emotional reactions to the situation or each other. The third category is probably the closest to the traditional understanding of

NPSA as summaries of utterances. (4) NPSA can also be part of a combination of communicated discourse and thought presentation categories all of which are used together to condense an event (usually a debate) into a dialogue-like interaction.

Dialogic NPSA

The first category—dialogic Narrator's Presentation of Speech Acts—includes NPSA that introduce two voices in a single occurrence. See example 4.3.

4.3 Penzias and Wilson called Dicke, who quickly confirmed that they had unintentionally tapped into the reverberation of the big bang.

GREENE 2011: 40

There are two speech acts (blue) presented here and two speakers (green), each of whom is given voice. However, this is one NPSA and not two because both speech acts are part of one speech activity—a phone call initiated by Penzias and Wilson. The act of calling prompted the confirmation; that is the exchanges are sequential and present a dialogue not isolated utterances.

While NPSA in 4.3 introduces a dialogic exchange, it does so purely through the summarizing properties of NPSA. There is no clear "projecting [of] the characters' different personalities and mutual relationships"—some of the requirements of dramatization suggested by Semino and Short (2004: 92). Example 4.3, I would say, has a minimal degree of dramatization since it introduces an interaction but does not necessarily project character personalities or reveal the nature of their relationship. At the same time, this exchange between Penzias and Wilson and Dicke is not unimportant or trivial. It is a crucial moment in the development of the narrative and a key event in the discovery process. Example 4.4 demonstrates an NPSA with a higher degree of dramatizing properties:

4.4 Miescher protested, but Hoppe-Seyler insisted on repeating the young man's experiments ... before allowing him to publish.

KEAN 2012: 20–21

In this exchange, the reader gets an indication of the kind of relationship between the two scientists: Miescher is clearly a less established figure who needs Hoppe-Seyler's permission to publish his findings. Hoppe-Seyler, on the other hand, is the dominant presence in the relationship. At the same time, Miescher's personality is such that he is capable of expressing dissatisfaction to his superior, while Hoppe-Seyler's character is revealed as somewhat more careful and not entirely accepting of a younger colleague's success. This is a good example of dramatization, and combined with the summarizing properties of NPSA, it allows the author to keep the narrative short without sacrificing characterization.

Strings of NPSA and Interaction of Discourse Presentation Types

Another way to use Narrator's Presentation of Speech Acts to introduce interactive exchanges into the narratives is to employ not one but a string of NPSA, which would form a backbone of an argument constructed using various types of presented discourse. Strings of NPSA are often used to introduce the readers to scientific debates, imagining them as immediate interactive exchanges rather than presenting them more accurately as a series of publications or conference presentations. Consider example 4.5.

4.5 (1) Most scientists in the mid-1960s explained the origin of mitochondrial DNA rather dully, arguing that cells must have loaned a bit of DNA out once and never gotten it back. But for two decades, beginning with her Ph.D. thesis in 1965, (2) Margulis pushed the idea that mitochondrial DNA was no mere curiosity … Margulis argued, a large microbe ingested a bug one afternoon long, long, ago, and something happened: nothing. Either the little Jonah fought off being digested, or his host staved off an internal coup … And after untold generations, this initially hostile encounter thawed into a cooperative venture. (3) Her opponents countered (correctly) that mitochondria don't work alone; they need chromosomal genes to function, so they're hardly independent. (4) Margulis parried, saying that after three billion years it's not surprising if many of the genes necessary for independent life have faded, until just the Cheshire Cat grin of the old mitochondrial genome remains today. (5) Her opponents didn't buy that—absence of evidence and all—but unlike, say, Miescher, who lacked much backbone for defending himself, (6) Margulis kept swinging back. (7) She lectured and wrote widely on her theory and delighted in rattling audiences. (She once opened a talk by asking, "Any real biologists here? Like molecular biologists?" She counted the raised hands and laughed, "Good. You're going to hate this.").

KEAN 2012: 103–104

The NPSA are in blue and create the frame for the debate [especially (1)–(5)], presenting it as an immediate exchange rather than as a prolonged argument that played out in scientific journals, as the remainder of the narrative and the reference to Margulis' Ph.D. thesis indicate it to be. It is worth noting that NPSA do not work alone to dramatize the debate. They are supplemented by Indirect Speech (in green), Free Indirect Speech (in red), Direct Speech and narration, more specifically, commentary from the author (in brown). The NPSA supply the outline with DS, IS, FIS, and author commentary providing the details and expanding the ideas introduced by NPSA. Each dialogic turn, in fact, begins with an NPSA.

While NPSA, DS, IS, FIS, and narration work together to dramatize the event, each type of communicated discourse presentation retains the properties prototypical for it in the narratives of discovery. Since NPSA are most often associated with the presentation of dialogue, they guide the exchange. IS, as I will show in greater detail in the following chapter, can be used to introduce scientific hypotheses, and this is exactly what both instances of IS do here. The hypothesis for the narrative is presented in the form of an argument to keep in line with the general concept of the scientific debate presented as a dialogue. The first occurrence of IS is introduced by the reporting verb "argued." The second occurrence uses a neutral reporting verb "saying"; however, it follows an NPSA that positions the hypothesis as a counter argument, "Margulis parried."

The verbs used in NPSA project confrontation, but they do so without clearly indicating if the argument played out in print or orally. This particular portion of a narrative is also a good example of how speech and writing presentation are treated as interchangeable. The mention of the thesis suggests that the initial argument was introduced in writing; however, the verb "pushed" is ambiguous in this regard. So are all the other verbs—"explained," "countered," "parried," "didn't buy," "swinging"—associated with the NPSA. All of them except "explained" are used in their metaphorical senses since their literal meanings indicate physical rather than verbal actions, and verbal actions associated with physical confrontation at that.

Metaphorical use of action verbs is not unique to discourse presentation in popular science, but there is a definite trend for such usage in popularizations. For instance, in the narratives of discovery, the focus is often placed on the intellectual actions of scientists rather than on the experimental procedures—one more way of showcasing the mental prowess of the members of the scientific community. This inevitably leads to a focus on presented thoughts even when verbs of physical activity are used, since they are, for the majority of the time, denoting intellectual actions. For example, in du Sautoy's (2011: 69) narrative about Crick and Watson's discovery of the shape of viruses, the actions of the scientists are mostly intellectual ones. Even when the verbs describing them denote physical activities such as "cracked," "looked," "found," "home in," they are used in their metaphorical senses to describe thought processes. At the same time, the narrative makes it clear that the discovery was made with the help of certain tools. The mentions of "the two-dimensional pictures from X-ray diffraction," "images," and "another diffraction image" suggest that the discovery was not purely an intellectual activity and that the scientists obtained their data as a result of an experiment. I will discuss thought presentation in connection with experiment descriptions in more detail in chapter 5.

To return to the analysis of example 4.5, as parts of NPSA the metaphorically used verbs embody the dramatizing properties because they present the argument in terms of a physical fight. However, coupled with the more traditional reporting verbs that introduce IS and DS ("saying," "asking"), they are to be interpreted as denoting verbal actions, more specifically speech. The fact that NPSA introduce aggression, allows for reporting verbs used in DS and IS to remain neutral.

Semino and Short's (2004) data reveals a similar use of physical action verbs; however, Semino and Short (2004) did not discuss this issue. While they also did not look at the verbs used in NPSA, their Appendices 3–6 list reporting verbs for direct and indirect speech and writing. There are 12 physical action verbs identified. Out of those, 9 are associated with non-fiction only. This tends to suggest that the metaphorical use of physical action verbs is more common in non-fiction, and the phenomena observed in the narratives of discovery are not random. Such use of the verbs in the NPSA also appears to be a careful choice made to dramatize the events.

Verbs in the NPSA are not the only examples of metaphors. The role of FIS in example 4.5 is to introduce a metaphorical explanation of the scientific hypothesis. This is a common use of IS and FIS in the narratives, where these forms of communicated discourse presentation often function as vehicles for creative explanations, which are attributed to the scientists (see chapter 5). Considering the first instance of FIS (in red), it is unclear if Margulis herself employed the Biblical reference, or whether it is coming from the author, but the use of FIS attributes the metaphor to the scientist. The same happens with the second occurrence of FIS. It also uses a metaphor that follows the hypothesis. It is possible to see these portions as narration since they appear out of place in a scientific debate. However, they are different from the instances of narration (in brown). Narration in this case tends to interrupt not explain. It is also evaluative and does not employ metaphors. In addition, there are other examples in the narratives of (Free)Indirect Speech that facilitate explanation through metaphors.

Direct Speech, like IS and FIS, is also used to expand the NPSA it follows. It shows an interaction outside the immediate dialogue ("a talk"), but the lecture

DS is connected with serves as an example and a continuation of the larger debate projected via NPSA.

Overall, as example 4.5 shows, dramatization can be achieved not through one or another single type of communicated discourse presentation but through a combination of several types. In the process, those types of communicated discourse that are not commonly associated with dialogue presentation (DS and IS) do contribute to dialogic exchanges. In fact, they play vital roles by introducing the subjects of the debates—the hypotheses—and supplying examples and explanations.

The relationship among the types of communicated discourse involved in dialogues, as in example 4.5, is such that each following type expands the ideas introduced in the preceding one. For instance, IS gives more details to the preceding NPSA. FIS that follows IS supplies explanations for IS. The level of emotionality (aggression, in case of example 4.5) decreases with each subsequent communicated discourse type. NPSA present the argument in the most aggressive manner using the verbs that describe a physical confrontation, while IS and DS use neutral reporting verbs, and FIS does not present arguments at all but provides explanations. Working together, the multiple types of communicated discourse produce a detailed dramatization of the events. At the same time, by making NPSA the backbone of the dialogic exchange, the author condenses a debate that took decades into a series of dynamic exchanges each of which is a reaction to a previous statement. This is an example of the summarizing and the dramatizing functions of NPSA working together and being supplemented by other forms of communicated discourse presentation.

Combining the various types of communicated discourse is not the only way of introducing interaction among the types of presented discourse. There are several examples in the corpus that use presentation of communicated discourse and presentation of thought to dramatize the narratives. Taking into account that presentation of thought alone is never used for dramatization, interaction between presentation of communicated discourse and presentation of thought deserves close attention.

Usually, studies of presented discourse focus on individual categories and discuss combinations of speech (writing) and thought in terms of contrast that can produce stylistic effects (see, for example, Toolan 2001, Short 2007). Semino and Short (2004: 153–159, 171–182) take a different approach. The relationship between speech/writing and thought presentation that they observe is hierarchical. To illustrate my point, I reproduce an example of an embedding from Semino and Short (2004: 172) as example 4.6. It shows an interaction between writing, thought, and speech presentation (original formatting preserved, [e] indicates embedding):

4.6 Example (18) is taken from Christopher Isherwood's autobiography, and is part of a long FDW [Free Direct Writing] quotation from his diaries:

(18) ⟨FDW⟩
[…] Thinking of Sister,
⟨eNRT level=1⟩
I remembered
⟨eIT level=1⟩
⟨eNRS level=2⟩
how I asked her, once
⟨eIS level=2⟩
what Vivekananda had been like.

In Semino and Short's (2004: 172) example, each following type of discourse presentation is structurally subordinated to the preceding one, with Free Direct Writing (FDW) being the subsuming type of presented discourse that includes others. Thus Indirect Speech is subordinated to Indirect Thought, and both are subordinated to FDW. Semino and Short's (2004) designation of levels further illustrates the hierarchical structure of the embedded interactions. Quotation phenomena—a stretch of presented discourse introduced as part of another presented discourse, as found in Semino and Short (2004), also follow a hierarchy with one type of presented discourse being the host to another.

The interactions between presentation of communicated discourse and presentation of thought in popular science work differently; when communicated discourse and thought presentation are combined, the relationship among the individual types of presented discourse remains linear. Consider example 4.7 (NPTA in green, IS in red, NPSA in blue, FIS in brown):

4.7 Rutherford thought long and hard about this strange result; it was, he remarked, as startling as if a bullet were to bounce off a sheet of tissue paper. Finally, at a Sunday dinner at his house in 1911, he announced to a few friends that he had hit on an explanation—that most of the mass of each atom resides in a tiny, massive nucleus.

FERRIS 1988: 256

In this example, each type of discourse presentation contributes to the narrative progression without an overarching category inside which other types exist. The transition from thought presentation to communicated discourse is smooth and logical. First, there was a thought (NPTA) that was later voiced (IS). Then a public announcement of the discovery was made (NPSA, FIS). It is possible to describe the relationship between presentation of thought and presentation of communicated discourse in this case as chronological. It is also worth noting that NPSA implies a dialogic exchange since it introduces Rutherford's audience.

On its own, the NPTA in 4.7 does not appear to contribute to dramatization. However, when it is seen in combination with communicated discourse, it becomes clear that the exchange introduced via communicated discourse (NPSA, FIS) would be impossible without the NPTA. In fact, it is possible to consider "he had hit on an explanation" as an NPTA embedded in an NPSA as a reminder of an earlier thought process. Regardless of one's position on the embedding, the NPTA contributes to dramatization when analyzed as part of the combination between presentation of thought and presentation of communicated discourse.

The example 4.7 is much more similar to what Toolan (2001: 122–123) and Short (2007: 231) include to illustrate a contrast between thought and speech presentation than it is to the interactions found in Semino and Short (2004). However, the instances of combination of communicated discourse and thought presentation appear to focus less on the contrast and more on a seamless transition from one type of presented discourse to another, preserving the chronological order of the events that led to the discovery.

The prominence of NPTA and NPSA in such combinations once again points to the importance of these forms of discourse presentation and highlights the dramatizing properties of NPSA. Every time NPSA are present, the combined stretch of discourse presentation has a degree of dramatization via dialogic exchange. NPTA are revealed as the most frequently used type of thought presentation to be associated with dramatization even if only in the presence of dialogic NPSA.

Borderline Dialogic NPSA

The third category of Narrator's Presentation of Speech Acts that contribute to dialogue creation and thus to dramatization are borderline dialogic NPSA. Examples of this category are used to set up an interactive exchange but give voice only to one interactant. Consider example 4.8:

4.8 Finally, at a Sunday dinner at his house in 1911, he announced to a few friends that he had hit on an explanation—that most of the mass of each atom resides in a tiny, massive nucleus.

FERRIS 1988: 256

The verb used in the NPSA in the example 4.8 suggests the presence of an audience, which is identified as "a few friends." However, these interlocutors are voiceless. Following the view of NPSA expressed by Leech and Short (1981/2007) and Toolan (2001), the omission of the audience's response is perhaps due to the unimportance and/or low relevancy of their remarks to the story. On the other hand, I suggest, NPSA may use silence as a form of non-verbal communication. In such a case, demonstrated in example 4.9 below, the voiceless interactant is still shown as reacting to the preceding voiced exchanges. The narrative from which example 4.9 is extracted talks about the Yang-Mills theory and relates an important point in its development—a seminar by Chen Ning Yang introducing the idea. During the seminar Yang was bombarded by questions from Wolfgang Pauli:

4.9 In the face of Pauli's onslaught, Yang found himself at a loss, and eventually he simply sat down quietly in the middle of his own seminar. Robert Oppenheimer, who was chairing the proceedings, coaxed him into resuming his talk, and Pauli stewed in silence.

CARROLL 2012: 155–156

There are two NPSA that show a loss of voice—"Yang found himself at a loss," and "Pauli stewed in silence." The reason I consider these occurrences to be NPSA is that they do not describe physical actions—Yang did not physically find himself, nor did Pauli really stew. The metaphorical use of the verbs, as I have noted above, is not unusual for NPSA. At the same time, even though no words are uttered, these are not thought acts either. It is possible to see these actions as behavioral processes (see Halliday and Matthiessen 2014: 214–215). However, the behavior here is clearly a loss of voice; this is why I suggest it would be more accurate to classify such occurrences as Narrator's Presentation of voiceless Speech Acts ($NPSA_v$), which constitutes a subcategory of NPSA. Rimmon-Kenan (2002: 63) gives a possible definition of "A character's speech" as "a silent activity of the mind," in addition to "conversation." These silences, I suggest, are employed strategically to communicate messages just as voiced speech acts do, and in example 4.9 they are responses to the voiced NPSA "Pauli's onslaught" and "coaxed him into resuming his talk." Being part of a communicative exchange, I suggest, allows these occurrences to be classified as NPSA and not narration.

The voiced NPSA in 4.9 are also good examples of the previous category of the dramatizing NPSA, where a string of two or more NPSA are used to construct an interaction. There are three speech acts in example 4.9: "Pauli's onslaught," "chairing the proceedings," and "coaxed him." Combined together they supply the skeleton of the interaction, with $NPSA_v$ introducing the details. The scheme of this interaction is shown in table 4.2, with each new line represent-

TABLE 4.2 *Schematic representation of interaction in example 4.9*

Pauli to Yang:	NPSA
Yang to Pauli:	NPSA_V
Oppenheimer to Yang:	NPSA, NPSA
Pauli to both:	NPSA_V

ing a reaction to a previous communicative move, as commonly found in a dialogue:

The extensive use of NPSA in the narratives of discovery in order to dramatize the events may be unexpected, yet it falls in line with the general preference of non-fiction for indirect forms of discourse presentation. The association between (Free)Direct Speech and faithfulness of representation (especially in non-fiction), I believe, is also a factor in favoring NPSA over (F)DS. Using (F)DS in a non-fiction text to present dialogue may create the false impression in the reader that the presentation is faithful and accurate when in reality it is dramatized. By not using (F)DS to present dialogue, the authors are giving themselves more license to fictionalize the events without being accused of misrepresentation. (For more reasons to use NPSA in popularizations see Appendix.)

The phenomenon of dialogue that is not a verbatim report of the anterior discourse is also common (according to Tannen 2007: 112 and Mildorf 2008: 289) in oral narratives that describe real-world events. There too, it is used as a dramatization mechanism that gives the speaker license to interpret and shape the material. Another reason for not using (F)DS might lie in the following argument introduced by Toolan (2001). Toolan (2001: 129) writes, "To opt for Direct Speech reporting is also to accept a scenic slowing of pace … [and] the inevitable fact that the narrated action will proceed … far more slowly." Using NPSA to present dialogue avoids this problem, as NPSA offer more authorial control and thus do not necessarily slow down the narrative pace.

Assigning the dramatizing function to NPSA goes against the established standard where indirect forms of speech and writing are considered more author-centered and therefore lacking the ability to present the events from the characters' points of view. NPSA have been traditionally associated with the summarizing function only (see, for example, Short 2007: 230). However, Short (2012) argues that the summarizing functions can be extended to all the categories of speech and writing presentation (presentation of communicated discourse, to use my terminology). The main point of Short's (2012) argument is as follows: NPSA summarize one proposition at a time (Short 2012: 24), while other types of communicated discourse presentation can be used to create summaries of "what was said rather than of a single proposition" (Short 2012: 25). The second type of summary Short (2012) calls a "discourse-domain summary." Short (2012: 24) warns that not all summaries are automatically NPSA and that researchers need to pay attention not only to the presence of a summary but also to what is summarized and the degree of summation.

To return to example 4.9, the last sentence demonstrates an example of both types of summaries: "Robert Oppenheimer, who was chairing the proceedings, coaxed him into resuming his talk, and Pauli stewed in silence." The first NPSA expressed via a relative clause—"who was chairing the proceedings"—would be a discourse-domain summary since it summarizes the discourse of chairing a physics seminar not one particular speech act associated with such an activity. The second NPSA, "coaxed him into resuming his talk" is a proposition-domain summary since it deals with one specific speech act and summarizes by eliminating the exact words used.

Short's (2012) extension of the summarizing properties beyond NPSA sets the precedent for regarding the functional scale for presentation of communicated discourse not as compartmentalized according to the categories but as fluid and gradual based on the degree of a function rather than on the binary presence/absence criteria. The fluidity of the functional scale for communicated discourse, I suggest, extends to the dramatizing properties as well as to the summarizing ones examined by Short (2012). The fact that (F)DS occurs just as often as NPSA supports the proposition that both of these categories share functions. Following Short's (2012) argument would suggest that they share summarizing properties. However, Short (2012: 25–26) and Semino and Short (2004: 95) show that the summarizing properties of (F)DS are employed to produce only one specific type of summary found primarily in newspapers—headlines. Since the narratives of discovery are not separate texts but fully-integrated segments inside popular science books, there are very few headlines to signal the beginnings of stories. Neither Short (2012) nor Semino and Short (2004) supply any other examples of summaries created via (F)DS. Short's earlier works (see, for example Short 1988, 2007, and Short et al. 2002) that deal with (but do not fully develop) the idea of the summarizing properties extending along the communicated discourse presentation scale also do not give examples other than headlines. Thus it is more likely that in the narratives of discovery (F)DS and NPSA share the dramatizing properties. Just like NPSA possess the higher degree of summarizing properties, so (F)DS does exhibit more of the dramatizing capabilities; however, this does not mean that NPSA cannot dramatize. The dramatizing function of NPSA, as demonstrated in popular science narratives of discovery, is not only present but quite pronounced, especially if the absence of (F)DS in the presentation of dialogue is taken into account.

The distribution and the degree of the dramatizing function in presentation of communicated discourse depends on the type of text analyzed. Thus in popular science narratives of discovery, NPSA possess a higher degree of dramatizing properties than they do in novels as analyzed by Semino and Short (2004), Leech and Short (1981/2007), and Short (2007).

Dramatization through (F)DS: Emotionality

Research into direct discourse presentation in non-fiction (see, for example, Bell 1991, Calsamiglia and Ferrero 2003, Moirand 2003, Semino and Short 2004, de Oliveira and Pagano 2006, Smirnova 2009, Urbanova 2012) singles out the following functions most commonly performed by (Free)Direct Speech: distancing the author from the presented claim, establishing credibility and reliability, supplying accuracy, and providing a personal perspective of the original speaker. In fiction, (F)DS is known mostly for its dramatizing potential. However, some studies (see, for example, Bell 1991: 209, Calsamiglia and Ferrero 2003: 169, Semino and Short 2004: 93–95) demonstrate that (F)DS retains its dramatizing properties even in non-fiction. In Bell's (1991: 209) words, (F)DS adds "a flavour" of the original speaker's "own words." It also, according to Semino and Short (2004: 95), may be used for "dramatizing protagonists' lives." When analyzing the functions of (F)DS in non-fiction, Bell (1991), Calsamiglia and Ferrero (2003), and Semino and Short (2004) draw on examples from newspapers and (auto)biographies. However, their findings on the dramatizing properties of (F)DS can be corroborated by the observations in popular science.

Because in the narratives of discovery (F)DS performs functions other than dramatizing (see chapter 5), it covers only some aspects of dramatization, namely emotionality and personal perspective, with presentation of dialogue reserved for NPSA. There are 26 occurrences of (F)DS that project emotionality and 12 occurrences that introduce the personal perspective of a scientist. The remaining 9 occurrences fulfill functions outside the realm of dramatization.

There are clear examples in the narratives of (F)DS being used to achieve emotionality and to infuse the narratives with "a flavour," to use Bell's (1991) word, of the scientists' speech. Consider examples 4.10 and 4.11.

4.10 "What a field of novelty is here opened to our conceptions!" he [Herschel] exclaimed, more delighted by the variety of the sky than bothered at having been wrong.

FERRIS 1988: 157

4.11 "Well boys, we've just been scooped," he [Dicke] told his colleagues as he hung up the phone.

BRYSON 2003: 12

Example 4.10 clearly shows Herschel's emotional state as he discovered that nebulae can be made of gas as well as of stars. Example 4.11 uses Direct Speech to expose the reader to Dicke's ways of speaking around his colleagues—the reader is to believe that Dicke referred to his fellow researchers as "boys" and used the verb "scooped" to indicate that Penzias and Wilson had beat them to the discovery of the cosmic microwave background radiation.

These examples also illustrate important differences in the presentation of (F)DS observed in the popular science narratives of discovery. For (F)DS that introduces emotions, the emotionality can be projected in two ways: either in the reported clause (projected as the emotionality of the scientists) or in the reporting clause (projected as the author's emotional evaluation). It is also possible to have both the reported and the reporting clauses contain emotionality. Example 4.10 shows emotionality that is projected as coming from both the scientist (use of an explicative and an exclamation mark) and the author (emotionally evaluative reporting clause—"he exclaimed"). There are 6 examples of this kind in the corpus, where the Narrator's Representation of Speech (reporting clause) is emotionally marked—what I label NRSe. When such a reporting clause is used in combination with emotionally charged DS, the author appears to reinforce the emotionality of the utterance. There are also 5 occurrences where emotion is projected as coming from the author only—emotionally charged reporting clause followed by neutral DS. Consider example 4.12 (NRSe is in blue):

4.12 Weber was so excited by the potential of their discovery that he prophetically declared, "When the globe is covered with a net of railroads and telegraph wires, this net will render services comparable to those of the nervous system in the human body, partly as a means of transport, partly as a means for the propagation of ideas and sensations with the speed of light."

DU SAUTOY 2011: 177

When Weber's direct speech is evaluated without du Sautoy's introduction, there is no expression of emotion. In fact, the reader would not know of Weber's excitement, nor would he/she necessarily feel excited about the information. However, the author's emotionally charged, evaluative reporting clause claims that the scientist produced this utterance as an exclamation of excitement upon the discovery of how to transmit messages via electric wires. The adverb "prophetically" reinforces the importance assigned to the discovery by Weber. In choosing to use an NRSe, du Sautoy is guiding the reader's emotional reaction to the DS, making sure that the reader understands not only the practical significance of science but also experiences the rise of emotions associated with discoveries. Reporting clauses in general (see, for example, Smirnova 2009) and in popular science texts in particular (see, for example, Calsamiglia and Ferrero 2003: 149, 156, 159; de Oliveira and Pagano 2006: 641, 644) are regarded as locations for authorial evaluation of presented discourse. The presence of NRSe in the narratives points to the author-controlled means of dramatization, just like dramatizing NPSA do. Calsamiglia and Ferrero (2003) note the influence of what they call "quoting frame" of an utterance presented via DS. They demonstrate using examples from Spanish newspapers describing and explaining the mad cow crisis that reporting verbs have the power to alter the readers' attitudes toward an issue and are often used this way by journalists. It appears that popular science authors also employ this technique.

It is noteworthy that the emotions NRSe project, and therefore the emotions that the reader is to experience, are always positive. They are emotions of excitement, enthusiasm, and, at times, surprise. Negative evaluation of discoveries, if introduced, always comes from the fellow scientists, not from the authors. Calsamiglia and Ferrero's (2003) findings suggest that by using reporting clauses the authors can either predispose or turn away the readers from the presented voices in the text. Clearly, the authors of the narratives of discovery want their readers to appreciate the scientists they write about and their achievements. This is one of the ways in which presented discourse contributes to the exclusively positive presentation of science and scientists.

There are also occurrences of DS that include neutral reporting clauses coupled with emotionally charged DS. Of these there are 10 out of 38 occurrences of DS. I should note at this point, that the majority of emotionality is projected through Direct Speech, with only 4 occurrences of Free Direct Speech (out of 9) that have emotionality markers. The combination of emotionally charged DS with a neutral reporting clause is more commonly associated with fiction where the characters are expected to speak for themselves and thus create an emotional attachment with the reader. For instance, Goffman (1981: 152) argues that "a full-scale story" requires that the narrator disengage for parts of the story from the immediate interaction with his/her addressee. Neutral reporting clauses in the narratives of discovery are evidence of such disengagement especially when analyzed against the examples of NRSe. Consider example 4.13.

4.13 Atomic scientist Ralph Lapp said, "I know what the other material is that the Argentines are using. It's baloney."
KAKU 2011: 236

In this example the neutral reporting clause ("Atomic scientist Ralph Lapp said") introduces a very emotionally charged statement. The reason for using a neutral reporting clause is two-fold. Firstly, there is no need to interpret Lapp's emotions for the reader; they are fairly clear. Secondly, by not evaluating the instance of DS, Kaku is turning a scientist into a character and is allowing Lapp to speak for himself. This utterance is clearly designed to expose the reader to the scientist's personality and to present his discourse as unedited—a feature of dramatization in fiction, according to Toolan (2001: 129–130).

Even though dramatization is not the main function of (F)DS in non-fiction (according to, for example, Semino and Short 2004: 93–95, Urbanova 2012: 51), in popular science, (F)DS that projects emotion outnumbers (F)DS that does not: 26 occurrences of (Free)Direct Speech are associated with emotionality while 21 perform other functions, some of which, presentation of personal perspective, for instance, are also associated with dramatization.

Dramatization through (F)DS: Personal Perspective

Presentation of personal perspective is a dramatizing facet of (Free)Direct Speech in the narratives of discovery. While (F)DS that introduces personal perspective is not as numerous as (F)DS that introduces emotionality, presentation of personal perspective is the second most frequent function of (F)DS in the corpus, with 12 occurrences. Even when (F)DS is not projecting emotionality, it is still most often used in a way that helps transform scientists from impersonal researchers into human characters. Example 4.14 shows Direct Speech used to present a personal perspective of a scientist:

4.14 … Lahn screened different populations alive today and determined that the brain-boosting versions [of microcephalin and aspm genes] appeared several times more often among Asians and Caucasians than among native Africans … follow up studies determined that people with these genes scored no better on IQ tests than those without them. Lahn … soon admitted, "On the scientific level, I am a little bit disappointed. But in the context of the social and political controversy, I am a little bit relieved."
KEAN 2012: 344–345

Lahn's DS is not necessarily emotional; its purpose is to show the scientist's personal reaction to the findings that have proven him wrong. Another example of DS being used to reflect a personal perspective is from the narrative about Yang and Pauli (discussed above in connection with NPSA and presentation of dialogue. See example 4.9 and table 4.2). A new extract from this narrative is introduced below as example 4.15.

4.15 The next day, Pauli sent a simple note to Yang: "I regret that you made it almost impossible for me to talk to you after the seminar. All good wishes. Sincerely, W. Pauli."

CARROLL 2012: 156

The narrative makes it clear that Pauli was quite obnoxious during Yang's seminar. As Carroll (2012: 155) tells the reader, "As an audience member in a scientific seminar, it may occasionally happen that you disagree with something the speaker is saying. The usual protocol is to ask a question, perhaps make a statement to register your disagreement, and then let the speaker continue. That wasn't Pauli's style. He interrupted Yang repeatedly ..." From this brief excursion into the etiquette of scientific seminars and the situation described in example 4.9, the reader knows that Pauli behaved discourteously. Yet, his personal message to Yang given in example 4.15 has no trace of remorse or apology; on the contrary, it places the blame on Yang, "you made it almost impossible for me to talk to you." Pauli's interpretation of the situation is radically different from that of the author and, presumably, from that of the other scientists involved. Using an instance of direct communicated discourse to relay Pauli's side is quite effective in presenting his perspective.

It is possible to interpret the use of DS in example 4.15 as distancing since the sentiment expressed appears not to be shared by the author. At the same time, when the narration around the utterance is considered, it becomes clear that DS does not distance the author from the scientist but is used by the author as almost a justification for the discussion of the flaws in Yang's theory. In general, the distancing function of (F)DS is very rare in the popular science narratives of discovery, and a more detailed exploration of the issue is offered in the following chapter.

Dramatization through IS and FIS: Emotionality and Dialogue

Unlike (Free)Direct Speech, Indirect Speech and Free Indirect Speech are not the common means of dramatization. Semino and Short (2004: 78) argue that IS "does not easily serve the purposes of dramatization." Of FIS Semino and Short (2004: 83–85) note that it can be used to create "irony at the expense of the person whose speech is being presented" and that FIS generally produces a distancing effect. Semino and Short (2004: 78) say that IS is more likely to occur in non-fiction (in newspapers especially). FIS, on the other hand, is more commonly associated with written fiction (see, for example, Rimmon-Kenan 2002: 116, Semino and Short 2004: 82). In fact, Rimmon-Kenan (2002: 116) argues that FIS has "a fictional ring even when found in other types of discourse." While this is an interesting proposition, the findings of my corpus do not support it, and FIS in the narratives of discovery, just like IS, has relatively unpronounced dramatizing properties.

The small degree of dramatization that is assigned to IS and FIS in the narratives of discovery limits its manifestation primarily to expressions of emotionality. There are 7 occurrences of IS (out of 34) that contain emotionality markers and 2 occurrences of emotional FIS (out of 4). Of the 7 occurrences of IS with emotionality, 6 contain emotionality markers in the reporting clause (NRSe), and 1 has an emotionality marker in the reported utterance. Examples 4.16 and 4.17 show emotionally marked IS and FIS respectively. The NRSe in 4.16 is in blue; FIS in 4.17 is in green with an emotionality marker in red.

4.16 ... light's speed, Einstein forcefully declared, is 300, 000 kilometers per second relative to anything.

GREENE 2011: 320

4.17 Far from rejoicing, the older scientist screwed up his brow and expressed his doubts that the nucleus contained any sort of special, non-proteinaceous substance. Miescher had made a mistake, surely.

KEAN 2012: 20

The emotionally marked reporting clause in example 4.16 (in blue) functions similarly to NRSe that accompany neutral DS. The reporting clause shows the reader the emotion that the author assigns to the scientist who produced the utterance. Since it is hard to determine whether or not a particular utterance was, in fact, accompanied by the emotion the author chose to associate with it, such use of IS fictionalizes the narratives and ejects positive perspective since the emotion projected through NRSe is designed to support the sentiment introduced through presented discourse of a scientist. Emotionally marked FIS functions similarly to emotionally marked DS: it attributes the emotional expression to the speaking scientists. However, the indirect form of discourse makes the effect less pronounced. In example 4.17 the emotionality marker contrasts FIS with the preceding Narrator's Presentation of Speech Act ("the older scientist ... expressed his doubts that the nucleus contained any sort of special, non-proteinaceous substance"), giving the reader a more direct sense of the scientist's emotional reaction.

When it comes to the presentation of dialogue, there is one instance when a dialogic exchange is created using IS. See example 4.18.

4.18 A member of the collaboration might say that the data is not yet ready for publication should never be used in a theoretical analysis. But a member of the audience might reply that data that isn't ready shouldn't be shown in public talks, either.

CARROLL 2012: 201

Note that IS in 4.18 is hypothetical—it presents not actual but possible utterances and does not attribute them to specific members of the scientific community. This kind of dialogue is purely fictional, and, as some analysts might suggest (for example, Skov Nielsen et al. 2015 a, b) the only possible kind of fictionality in non-fiction. As I have been demonstrating, however, fictionality in the narratives of discovery does not have to be tied to hypothetical utterances only. This kind of dramatization is different from the dramatization created via NPSA, where real, not hypothetical events, are played out as dialogic exchanges. While this example seems out of place compared with other instances of dramatization in the corpus, it does point to an important function of IS in popular science—that is the ability to introduce scientific hypotheses. A discussion of IS as a vehicle for scientific hypotheses and of other non-dramatizing properties of communicated discourse presentation is offered in the following chapter. (Practical applications of these functions are addressed in the Appendix)

Conclusion

This chapter began with a proposition that the function scale for presented speech/writing based on examples from fiction is not fully representative of the way communication of presented discourse functions in non-fiction, specifically, in popular science. For instance, the traditional scale denies the indirect forms of discourse (NPSA especially) any reader-engagement properties. Narrator's Presentation of Speech Acts is usually assigned the distancing and summarizing functions (see, for example, Short 2007), while dramatization is reserved primarily for (Free)Direct Speech.

With non-fiction, the situation is somewhat different and less clear since no function scale based solely on examples from non-fiction exists, and the functions of communicated discourse in non-fiction are usually evaluated in terms of the scale created for fiction (see, for example, Semino and Short 2004). However, those who research presented discourse in non-fiction (written or oral) note that the functions are not as neatly arranged as the scale created for fiction suggests (see, for example, Myers 1999; Short et al. 2002; Short 2012). In non-fiction, there is a fair amount of cross-over between the functions traditionally associated with direct and indirect discourse. For instance, (F)DS may perform both the distancing and the dramatizing functions (see, for example, Myers 1999, Calsamiglia and Ferrero 2003, Semino and Short 2004). In some instances, (F)DS in non-fiction possesses summarizing properties (see, for example, Short 1988, 2012, and Short et al. 2002). In fact, Short (2012) argues that any type of Public Discourse presentation can be used to create summaries. Thus his is the first study to suggest the possibility of universal functions for presentation of communicated discourse.

In my opinion, studies of discourse presentation in non-fiction have enriched the outlook on discourse presentation in general. However, only one genre has achieved most of the attention, and the newspaper article remains the best-analyzed non-fiction text. The data from a popular science corpus introduced here contributes to the discussion and reveals functions of presented discourse not previously observed in news reports and other non-fiction corpora.

This chapter extends Short's (2012) argument to include not only the summarizing function but the dramatizing function also as a universal function of communicated discourse. As this chapter demonstrates, NPSA (and to a lesser degree IS and FIS) can be the means of dramatization and reader-engagement in the narratives of discovery. NPSA are overall very prominent in the narratives of discovery.

The prominence of NPSA in non-fiction is not surprising. Semino and Short (2004), for example, note that this was the second most frequent category to occur in their corpus as a whole, and the frequency information shown in chapter 3 demonstrates that in the Semino and Short (2004) non-fiction section of the corpus, NPSA were the most frequent category. In general, it is expected that non-fiction favors indirect discourse. Presented discourse in non-fiction is generally regarded as having less to do with character creation and more with the construction of the author's professional identity. That means that more reformulation and interpretation would be necessary, activities which produce indirect discourse. The use of indirect discourse is expected of a genre like popular science where the authors have to introduce complex ideas in ways which are understandable for non-specialists. In this communicative situation, reformulation becomes key, as Ciapuscio (2003) demonstrates.

At the same time, popular science books are not as homogenous as they are believed to be. These texts are usually discussed in terms of their overarching function of explaining science to the public (see, for example, Turney 2004a,b), but functions of their individual components, like stories of discoveries, are largely ignored. The narratives of discovery function somewhat differently from the segments designed to explain the most cutting-edge theories. While explanation remains a vital goal of the narratives, they also connect the public with the scientific community by other means. One of them, I suggest, is presentation of scientists as characters in fiction with whom the reader can connect on an emotional level. In fiction, creation of such characters is usually achieved, in terms of discourse presentation, with (F)DS. However, in the popular science narratives of discovery, NPSA and, to a lesser degree, some other indirect forms of communicated discourse take on the character creating properties such as presentation of dialogue and showing of the relationships between characters.

The extension of the dramatizing and fictionalizing properties to indirect discourse likely happens because of the pressure of the conventions to use indirect discourse in non-fiction and because the narratives are fictionalzing real events rather than inventing new worlds and characters. (F)DS in addition to being an excellent means of dramatization is also associated with an expectation of accuracy [for a discussion of accuracy in (F)DS see Short et al. 2002 and Short 2012]. A lot of dialogue that is introduced in the narratives of discovery appears to be fabricated rather than reproduced from original interactions. For example, the authors tend to dramatize scientific debates that took place over a period of time as dialogue. Using (F)DS in such situations, where the characters and the issues are real but the interaction is fictionalized, might pose a problem. Using indirect discourse, and the most author-controlled type of it, avoids this predicament. NPSA allow the authors more creative freedom than (F)DS would.

At the same time, not all forms of indirect discourse contribute to dramatization in such an important way as NPSA do. It appears that IS and FIS possess some dramatizing properties but do not capitalize on them. These types of communicated discourse presentation exemplify how the dramatizing function can extend to all the categories of discourse presentation without eclipsing the prototypical uses of each discourse presentation type.

CHAPTER 5

Beyond Dramatization

Introduction

In the previous chapter I addressed the uses of communicated discourse presentation and presentation of thought from the point of view of dramatization and demonstrated that it is primarily communicated discourse that dramatizes the narratives by using Narrator's Presentation of Speech Acts and (Free)Direct Speech. I showed that presentation of thought and certain types of communicated discourse do not have much to do with dramatization. This finding raises the question about the primary roles of thought presentation, Indirect Speech, Free Indirect Speech, and non-dramatizing Narrator's Presentation of Speech Acts and (Free)Direct Speech. In this chapter, I discuss the functions of thought presentation and spend more time on non-dramatizing communicated discourse. I conclude that next to dramatization, the other major role of presented discourse in the narratives of discovery is introduction and explanation of scientific concepts and hypotheses. While explanation is a commonly observed role of non-dramatizing communicated discourse in popular science (see, for example, Moirand 2003), presentation of scientific hypotheses, and especially their introduction via presentation of thought is a less discussed phenomenon (see, for example, Pilkington 2018). In general, this chapter demonstrates that presentation of thought is more science-oriented than communicated discourse presentation, which tends to highlight personal relationships and arguments as examples in chapter 4 demonstrate.

The different roles performed by communicated discourse and presentation of thought and the different effects associated with each are not unexpected taking into account the juxtaposition of speech/writing with thought presentation highlighted in the previous studies of presented discourse (see below for examples). In this chapter, however, I will focus not only on the differences but also outline a number of connections between presentation of communicated discourse and presentation of thought.

Those who study speech/writing and thought presentation often note the functional distinctions. For instance, Semino and Short (2004: 118) confirm that "the effects that result from their [thought presentation] types are quite different from those we have noted for speech and writing." The difference, as Leech and Short (2007: 270) point out, lies in the inaccessibility of thought, "We cannot see inside the minds of other people." Short's later works (see, for

© KONINKLIJKE BRILL NV, LEIDEN, 2018 | DOI: 10.1163/9789004365971_006

example, Short 2007 and Short 2012) further this point by stressing that thought presentation does not possess those communicative properties that are common for speech/writing presentation. In addition, Short (2007: 231, 2012: 23) notes that thought presentation categories do not have summarizing properties: Narrator's Presentation of Thought Acts do not really present summaries of propositions but rather indicate "the extent of narrator 'interference'" compared to Free Direct Thought, for example (Short 2007: 231).

It is commonly assumed that since thought is not directly accessible in everyday life and is not used for communicative purposes, presentation of thought centers on the inner worlds of actants (see, for example, Toolan 2001, Semino and Short 2004, Leech and Short 2007, Short 2007, Short 2012). As a result, thought presentation is usually discussed in close connection with dramatization. While my corpus follows the observations of Semino and Short (2004) regarding the frequencies of Direct Thought, Free Direct Thought, and Free Indirect Thought (see chapter 3), the functions associated with presentation of thought are radically different from those assigned to thought presentation by the previous studies, which dealt with non-fiction genres other than popular science. By using presentation of thought to introduce scientific hypotheses and discoveries, the authors of the popular science narratives take thought presentation out of the realm of the intimate. Presentation of thought introduces the reader to common knowledge or cutting edge scientific advancements rather than to the intimate thoughts of the scientists. In other words, the messages delivered through presentation of thought are very much public and, for the most part, do not contribute to characterization of the scientists, something that would be expected of thought presentation based on the previous analyses.

Presentation of thought in the narratives of discovery is closely associated with the experimental and the empirical (and thus the more physical rather than the mental) sides of science. The decision to focus on the empirical side of science through the presentation of mental processes might be an attempt at establishing a kind of closeness between the reader and the scientific issues discussed.

In the narratives of discovery overall, experimental procedures are not foregrounded, and the discoveries are often described as being the products of intellectual rather than experimental processes, so it appears that overall scientific discovery is presented as a result of thinking rather than doing (see discussion of metaphorical use of verbs as part of the analysis of example 4.5 in chapter 4). Presented discourse brings the experimental procedures to the forefront even if they are introduced through presentation of thought and, therefore, internal modes of discourse presentation. While narration emphasizes the theoretical and the intellectual side of the discovery process, presented thoughts of scientists are more strongly aligned with experimentation. The authors do not always take it upon themselves to give accounts of experiments but prefer to relate this information through presented discourse of the scientists who performed the experimental procedures, thus creating more authentic accounts.

This chapter is devoted to the discussion of thoughts of scientists associated with the presentation of scientific hypotheses and discoveries as well as to an exploration of non-dramatizing communicated discourse. Even those forms of communicated discourse that are associated primarily with dramatization [NPSA and (F)DS] can contribute to the presentation of scientific ideas. To carry over an argument from the previous chapter, individual types of presented discourse do not have to be confined to single functions. For example, while the primary role of (F)DS in the narratives may be dramatization, this type of communicated discourse presentation has other uses as well. However, they do not align with the traditional distancing functions of (F)DS common in non-fiction. In fact, (F)DS is more likely to convey the alignment of the author and the scientists, in some cases to the point that the presented voices are allowed to take over the narratives—with FDS significantly contributing to narrative progression and fulfilling the functions of narration. (F)DS is also a popular means of including explanations.

Non-dramatizing NPSA also play their part in introducing scientific ideas. However, these ideas are quite often supplemented by explanations provided by other forms of presented discourse or narration.

Hypotheses and Discoveries: Presentation of Thought is Not so Private

Presentation of thought in the narratives of discovery is dedicated not so much to the unveiling of the inner worlds of scientists but rather to tracing the mental processes and reactions to empirical work which result in discoveries. Narrator's Presentation of Thought Acts and Indirect Thought aid the most in accomplishing this goal. These two types of thought presentation are responsible for introductions of scientific hypotheses and discoveries in the narratives.

IT and NPTA are the least dramatic means of thought presentation. For instance, Toolan (2001: 139) writes, "... recourse to more direct thought-presentation than IT [FIT, DT, FDT] may ... invite the inference of ... 'entering' of the character's intimate mental space." Like Toolan (2001), Semino and Short (2004: 128, 131) observe that IT and NPTA are "less dramatic" means of accessing the inner world of a character. Semino and Short (2004: 115) note the relative lack of NPTA across their corpus and in the non-fiction section in particular. It is IT that is more prominent in their frequency counts (Semino and Short 2004: 115). In my corpus, however, both IT and NPTA have nearly the same number of occurrences, accounting for almost all of the instances of thought presentation (see chapter 3).

The preference for the most non-dramatizing categories of thought presentation falls in line with the purpose of IT and NPTA in the narratives of discovery, where presentation of thought focuses on science rather than on the scientists. Out of 53 occurrences of thought presentation, 44 focus on the scientific issues rather than present more personal thoughts.

The introductions of scientific hypotheses and discoveries reveal several important points about presentation of thought in popular science. Firstly, they demonstrate a strong connection between presentation of thought and narration that describes experimental procedures, thus supplying evidence for presentation of thought being concerned more with the empirical than with the intimate. Secondly, presentations of discoveries and hypotheses follow specific verb patterns that generalize presentation of thought and, again, indicate that this type of presented discourse does not focus on expressions of individual inner worlds. In fact, I suggest that verb choices used in reporting clauses of thought presentation function as what Mildorf (2008: 288) calls "a mitigating strategy which helps the speaker disclaim any ultimate knowledge or access to … other people's minds." Mildorf's (2008) study is of oral narratives, and consequently the mitigating strategies for thought presentation she observed are different; however, her analysis is valuable for identifying the phenomenon of a mitigating strategy, which, as my findings demonstrate, could extend beyond spoken discourse.

Other than emphasizing scientific issues, presentation of thought is also responsible for contributing to a positive image of scientists through highlighting hypotheses that have been proven correct and showing the scientists as creative thinkers when it comes to descriptions of their discoveries.

I will begin the discussion of thought presentation in the narratives with an examination of hypotheses and follow up with an exploration of discovery descriptions.

Scientific Hypotheses: NPTA and IT

Scientific hypotheses are an important part of the narratives of discovery. They are signposts that guide the readers' expectations of what is to be discussed in a narrative. Even though only about 20% of the narratives use presented discourse to introduce hypotheses, these narratives rely overwhelmingly on presentation of thought to do so. Out of all the occurrences of thought presentation, 43.3% are used to present hypotheses, with only 13.5% of communicated discourse occurrences (mostly IS) serving the same role. Example 5.1 shows a hypothesis presented via Indirect Thought:

5.1 Many scientists at the time were skeptical, but Shope wondered if rabbit "horns" were also tumors, somehow triggered by an unknown virus.
ZIMMER 2011: 24

This is a typical hypothesis expressed using presentation of thought. It contains a verb of mental action ("wondered") and a hedge ("if") to indicate uncertainty that is to be eliminated once the proof is obtained. Using presentation of thought to introduce a hypothesis inevitably personalizes the discovery process.

There are two major ways to introduce scientific hypotheses using NPTA and IT. The first one involves interaction of communicated discourse, thought presentation, and narration and involves mostly NPTA. The second one is to use single occurrences of IT, as in example 5.1.

Narrator's Presentation of Thought Acts that appear in combinations with presentation of communicated discourse and narration do not themselves introduce hypotheses and are more likely to resemble prototypical NPTA observed by Semino and Short (2004: 130) and described as "occurrences of a specific individual thought in the mind of a participant in the story, which do not include any indication of the propositional content or the 'wording' of the thought." Such NPTA, as Semino and Short (2004: 131) suggest, most often introduce the character's motivations that help explain his/her speech or actions that precede or follow. This is also true for this category of NPTA in my corpus. However, the thoughts and motivations that they present are connected with scientific hypotheses and never uncover personal feelings of the scientists. Consider example 5.2 (NPTA is in blue followed by presentation of communicated discourse in green):

5.2 While delving microscopically through the pus in surgical bandages, Miescher found a substance he didn't recognize and called it nuclein (because it resided in the nuclei of cells). At the time, Miescher did little more than note its existence, but nuclein clearly remained on his mind, for twenty three years later in a letter to his uncle he raised the possibility that such molecules could be the agents behind heredity.
BRYSON 2003: 400

As Semino and Short (2004: 131) note, NPTA of this kind often appear as interjections within characters' conversations. While this is not the case in my corpus, NPTA that do not themselves express hypotheses are much more likely to be followed by presentation of communicated discourse than the other types of thought presentation. For instance there are 17 prototypical NPTA and 5 of them are followed by presentation of communicated discourse. As with example 5.2, communicated discourse in such combinations is motivated by NPTA and contains the hypothesis. In a way, NPTA of this kind, while not introducing the hypotheses themselves, trace the mental path the scientists take to arrive at a hypothesis.

There is also a connection of communicated discourse presentation and presentation of thought with narration that such NPTA highlight. Note that in example 5.2, the narrated segment (unmarked) describes the experimental procedure. Example 5.3 (NPTA in blue, IS in green, narration unmarked) also demonstrates the interaction of thought presentation with the experimental procedure presented via narration. The IS that follows the NPTA contains the hypothesis, to which the narration and the NPTA lead. The rest of the narrative discusses and confirms this hypothesis.

5.3 Thompson and his colleagues at the Cavendish Laboratories began to measure the electrical charge and the weight of some of these radiations. They tried to decide how these two measurements were related to each other. In 1987 Thompson proposed that these rays were streams of charged subatomic particles: bits of atoms.

BYNUM 2012: 183

Just as with the other instances of discourse interactions addressed in chapter 4, the different types of discourse presentation introduce the different chronological stages of the discovery process.

Hypothesis-introducing NPTA may also manifest as single occurrences (there are 9 such cases) that themselves carry the hypotheses. Consider example 5.4.

5.4 This set Bunsen to wondering whether they might be able to detect chemical elements in the spectrum of the sun as well.

FERRIS 1988: 164

Single occurrences of hypotheses presentation are, however, much more commonly expressed via IT (see example 5.1). In fact, there are only 4 interactions between presentation of communicated discourse and thought presentation that include IT. Single-occurrence hypotheses presentations reveal clear verb choices associated with hypotheses expressed through NPTA and IT. Consider example 5.5.

5.5 He wondered if something other than bacteria was responsible for tobacco mosaic disease, something far smaller.
ZIMMER 2011: 4

The hypothesis is introduced via IT. The verb "wondered," is part of a pattern for hypotheses presentation via thoughts of scientists. There are four verbs/verbal phrases that occur most often with hypotheses in presentation of thought: "wonder," "come up with an idea," "think," and "assume." The first two most often indicate what I call "positive hypotheses"—hypotheses that are proven correct as the narrative progresses. The last two are associated with "negative hypotheses"—hypotheses that are later refuted. Accordingly, example 5.5 introduces a positive hypothesis, while example 5.6 illustrates a negative hypothesis:

5.6 Perhaps, he thought, the plants were suffering from an invisible infection.
ZIMMER 2012: 3

Other than the verb used to present it, it is impossible to distinguish a negative hypothesis from a positive one without the context of the narrative. The verb choice, however, creates a distinction between the two hypotheses and can help predict the narrative's resolution when hypotheses are analyzed in isolation. Examples 5.5 and 5.6 are taken from the same narrative on the discovery of viruses and show that other than the verb used in the reporting clause of Indirect Thought, there is nothing to separate a negative hypothesis from a positive one. For instance, both contain hedges ("if" and "perhaps") that point to the tentativeness of the statements. This demonstrates the key role reporting verbs play in presentation of thought introducing hypotheses.

I mentioned at the beginning of the chapter that while presentation of thought dominates introduction of hypotheses over presentation of communicated discourse, Indirect Speech also contributes to presentation of hypotheses. The numbers are, however, lower: only 14 hypotheses presented via IS compared to 26 introduced with the aid of thought presentation. There are certain verbs ("argue" and "propose," for example) which are frequently used in IS that expresses hypotheses, but the verb choices pointing to positive and negative hypotheses are less pronounced. While the verbs "propose" and "suggest" are associated with positive hypotheses, there is no discernible verb pattern for negative hypotheses. Both positive and negative hypotheses could be expressed using the verbs "argue" and "ask." Consider examples 5.7 and 5.8.

5.7 Everett argued that Schrodinger's equation should apply to everything because all things material, regardless of size, are made from molecules, atoms, and subatomic particles.

GREENE 2011: 321

5.8 He argued that this first cell obviously contained a complete set of molecular instructions, but that each time the zygote and its daughter cells divided, the cells lost half of their instruction.

KEAN 2012: 130

It is hard to distinguish the positive from the negative hypothesis in this case. Example 5.8 uses an intensifier "obviously" which projects certainty, while 5.7 hedges the proposition with the use of "should." Based on this observation, it is possible to assume that example 5.7 presents a negative hypothesis and 5.8 a positive one when, in fact, the opposite is true.

The lower number of hypotheses introduced via IS and the less rigorous verb choices point to the dominance of thought presentation when this particular function is concerned. The preference for presentation of thought suggests the focus on the personal observations and the empirical work on the road to discovery rather than on publications, for example, which would manifest as instances of communicated discourse presentation.

In both the hypotheses introduced via presentation of thought and via IS, positive hypotheses outnumber the negative. The emphasis on the positive hypotheses also confirms the general focus of the narratives on the positive outcomes of science. Already the majority of the narratives tell the stories of successful discoveries, while very few narratives describe what I call "failed discoveries."

The narratives of failed discoveries show what happens when the process of scientific discovery fails: either the ideas do not find the experimental support, or the experimental results are not recognized as significant. Examples would be stories of the cold fusion discovery by Pons and Fleischmann (see, for instance, Kaku 2011: 236–237) or accounts of how "The Babylonians came close to discovering a very important number in mathematics: 0" but did not quite manage to "think of 0 as a number in its own right" (du Sautoy 2011: 23). In such narratives, the break down can happen on either the experimental (cold fusion narrative) or the conceptual (the Zero narrative) level. In both cases, the narrative pattern will continue to "recycle" until a positive outcome or a positive evaluation can be obtained (Hoey 2001: 130–133). (For more on the structure of narratives of failed discoveries see Appendix.) As examples of dissatisfaction with a failure and the repeated attempts to correct the problem or to justify it, narratives of failed discoveries are another indicator of the overall celebratory nature of popular science and its determination to present science and scientists as infallible. Negative hypotheses, however, are much more likely to appear not as focal points in the narratives of failed discoveries but as side steps in the narratives of success.

The prevalence of positive hypotheses is only one example of presented discourse being used to boost the image of scientists. Another way presentation of thought contributes to presenting scientists in a positive way is by showing them as creative thinkers. Descriptions of discoveries incorporating figurative language serve as examples, and I discuss them later in the chapter.

Introduction of Discoveries: IT

After hypotheses presentation, the introduction of discoveries forms the second most numerous functional category of thought presentation. There are 12 occurrences of thought presentation that introduce discoveries, and all of them use IT. Indirect Thought appears to be the only mechanism of presented thought for signaling these important points in the narratives. (Other options include narration and non-dramatizing NPSA.) Consider example 5.9.

5.9 An atom, Rutherford realized, was mostly empty space, with a very dense nucleus at the center.
BRYSON 2003: 140

This example demonstrates the most popular way of incorporating statements of discoveries into the narratives: the use of what I label "pragmatic" as opposed to figurative language. Such descriptions use generic terms of the disciplines to present the discoveries. This method of discovery presentation reveals the connection of IT with presentation of communicated discourse and narration and also uncovers a popular verb choice associated with discovery presentation. In other words, Indirect Thought that presents discoveries acts similarly to hypotheses-presenting Narrator's Presentation of Thought Acts.

Almost all of the discoveries introduced via IT use the verb "realize"—10 out of 12 total occurrences. The remaining two instances of IT that introduce discoveries use the verbs "occur" (see example 5.12 below) and "assume." Except for the instance that uses "assume" ["she assumed ... that the change involved hydrogens shifting around" (Kean 2012: 100)], all of the discoveries introduced through IT contain an element of sudden enlightenment, which the verbs "realize" and "occur" express. At a first glance this observation suggests that some scientific discoveries are being presented as serendipitous insight not necessarily dependent on consistent empirical work. However, when the instances of discoveries introduced via IT are examined in the context of the narratives, the connection between presentation of thought and descriptions of experiments provided via narration show that just the opposite is true. Out of the 12 occurrences of discovery introductions via IT, 8 are preceded and/or followed by descriptions of experimental procedures, presenting the eureka moment as a reaction to a specific experimental result. Consider example 5.10, which supplies the rest of the narrative introduced in the previous example:

5.10 In 1910, Rutherford ... fired ionized helium atoms, or alpha particles, at a sheet of gold foil. To Rutherford's astonishment some of the particles bounced back. It was as if he said, he had fired a fifteen-inch shell at a sheet of paper and it rebounded into his lap. This was just not supposed to happen. After considerable reflection he realized there could be only one possible explanation: the particles that bounced back were striking something small and dense at the heart of the atom, while the other particles sailed through unimpeded. An atom, Rutherford realized, was mostly empty space, with a very dense nucleus at the center. This was a most gratifying discovery, but it presented one immediate problem. By all the laws of conventional physics, atoms shouldn't therefore exist.

BRYSON 2003: 139–140

With such an introduction, the discovery presented as IT does not seem as sudden or as unfounded as it might when IT is analyzed in isolation as in example 5.9. Note also the interaction of narration (red), communicated discourse: Indirect Speech and Free Indirect Speech (green), and presentation of thought: Indirect Thought (blue). The narrated segment supplies the details of the experiment (and later the evaluation of the discovery); IS and FIS show Rutherford's reaction to the experiment, and the second sentence of IT presents the discovery itself as a culmination of all the previous activities provided via narration and presentation of communicated discourse. The combination of presented discourse (communicated discourse and presentation of thought) and narration works to create a chronological account of the discovery and to position it as the outcome of the experiment. Rutherfords' thoughts are presented as focused on the observation, from which he deduces the structure of the atom—his discovery.

When IT is not connected with an experimental procedure, the discovery process appears underdeveloped and the discovery announcement comes somewhat suddenly. However, this mode of introducing discoveries appears to be a deliberate choice on the part of the authors. A discovery presented through IT which is not connected to an experiment is usually not the main discovery of a narrative (see example 5.11), or it mimics the actual events where a discovery was indeed a sudden realization (see example 5.12).

5.11 Despite the evidence mounting against him, Bekenstein had one tantalizing result on his side. In 1971, Stephen Hawking realized that black holes obey a curious law.

GREENE 2011: 247

In this case, Hawking's discovery helps Bekenstein strengthen his own theory and provides evidence for the main discovery of the narrative—the multiverse. There is no need to take up room with the descriptions of Hawking's discovery process; the mention of the discovery is sufficient. The remainder of the narrative provides a brief explanation of Hawking's findings and connects them to Bekenstein's ideas.

5.12 As often happens, the answer came to him not while he was at work in his observatory but while he was relaxing. While on a boat in the Thames, Bradley found himself gazing at a wind vane mounted atop the mast. It pointed into the wind and therefore seemed to change direction whenever the boat turned. What was changing, of course, was the orientation, not of the wind, but of the boat.

It occurred to Bradley that the earth is like a boat adrift in winds of starlight—that, as the earth moves through the starlight, its motion alters the apparent positions of the stars.

FERRIS 1988: 138

Example 5.12 presents a different reason for not including a description of the experiment—it did not happen—and illustrates the use of figurative language in the description of the discovery. Ferris (1988: 137–138) tells a discovery story different from those found in the majority of the narratives but one that nonetheless is not uncommon when it comes to discoveries, as he himself suggests.

Thought presentation is usually regarded as arbitrary since it is, most of the time, impossible to connect it to the anterior discourse (see, for example, Short 2012: 23). That means that the criterion of faithfulness is even less applicable to thought presentation than it is to speech/writing. This is what leads some researchers (see, for example, Cohn 1990: 784–785, Dawson 2015: 80) to suggest that if at any time presentation of thought occurs without specific references to memoirs, journals, or similar materials, it therefore contributes to fictionality. This line of argument gains more strength once presentation of thought becomes detailed and includes figurative language as is the case with example 5.12. The analogy between the boat on the water and the earth in space is clearly attributed to Bradley, but it is not clear if the scientist expressed it in quite the same creative manner as Ferris (1988: 138) does. In the last sentence of example 5.12, is the reader experiencing the creativity of the writer or that of the scientist?

With IT, the content rather than the wording is in the forefront; however, analogies and metaphors shift the focus back to wording, producing IT that is more suitable for fiction, where the narrator has unlimited access to the thoughts of his/her characters. Incidentally, example 5.12 does not employ the verb "realize" commonly associated with the introduction of discoveries via IT and which I consider a mitigating mechanism which signals the generalized and non-intimate nature of thought presentation. By using the verb "occur" Ferris (1988: 138), in a way, breaks with the tradition and thus signals an instance of presented thought similar to thought presentation usually associated with fiction.

While I do not subscribe to the line of argument that suggests any manifestation of presented thought automatically designates a text as a fiction, I still believe that more elaborate instances of presented thought (as in example 5.12) contribute to dramatization and thus do fictionalize the texts in which they appear. In general terms, thought presentation is usually discussed in close connection with dramatization. For instance, Toolan (2001: 139) and Semino and Short (2004: 123, 128, 131) explain the differences in the effects produced by thought presentation categories in relation to the degree of dramatization. Semino and Short (2004: 121, 123) also note that there is not much difference in the effects produced by thought presentation in fiction and non-fiction: in both thought presentation is connected with dramatization, differing only in the degree. Thus Direct Thought, Free Direct Thought and Free Indirect Thought (the types of thought presentation considered as possessing the highest dramatizing properties by Toolan and Semino and Short) are more common in fiction, while almost nonexistent in non-fiction.

The scarcity of these forms of thought presentation in the narratives of discovery (see chapter 3) supports the claim made in the beginning of this chapter that the functions of thought presentation in popular science narratives of discovery lie beyond dramatization. Example 5.12 is the only instance of presented thought that uses figurative language. The overwhelming majority of occurrences express the thoughts of scientists in more pragmatic language less likely to raise questions of faithfulness. I suggest that the lack of elaborate descriptions in presentation of thought and the preference for reporting verb patterns (certain verbs associated with positive and negative hypotheses and almost universal use of the verb "realize" for the presentations of discoveries) constitute the mitigating strategies for presentation of thought that point to the overall lack of inner world access in the narratives.

At the same time, the incorporation of figurative language in such a way that it is attributable to scientists shows the willingness of the authors to present the scientists as capable of creative approaches to their work. Figurative language is, however, not common for presentation of thought, but it is more likely to occur in non-dramatizing Indirect Speech.

Non-dramatizing Communicated Discourse: Explanation of Science

Indirect Speech

I will begin the discussion of non-dramatizing communicated discourse with an exploration of Indirect Speech since its non-dramatizing functions relate closely to those of thought presentation. Indirect Speech is the only type of communicated discourse to introduce scientific hypotheses and is the most likely form of presented discourse to include figurative language.

Traditionally, IS, just like Narrator's Presentation of Speech Acts, has been valued for its ability to present the message but omit its exact wording, which is potentially a useful aspect when it comes to popularization since not all discourse produced by scientists can be clear and straightforward enough to be incorporated via (Free)Direct Speech. Some of it, as Ciapuscio's (2003) study suggests, may be too technically worded to be quoted directly and requires reformulation. IS supplies an excellent means of reformulation. In fact, the authors in the corpus sometimes use IS to infuse the narratives with figurative language. I have identified three types of tropes found in IS: prosopopoeia,[1] metaphor, and analogy.

The decision to include figurative language as part of IS can make the reader wonder whether it is the author or the scientist who is responsible for the figurative language. This is especially relevant for metaphors and analogies since

1 Prosopopoeia is a rhetorical device that presents a speaker's voice as coming from an inanimate object or from another person. I consider written works or theories treated as speakers in the narratives examples of prosopopoeia.

they are parts of IS itself rather than of the reporting clause as is the case with prosopopoeia.

As Ciapuscio (2003: 209–210) notes, "science popularization texts originate in other texts," which often include professional research articles. Using prosopopoeia the authors can emphasize this connection in a somewhat unusual way: written works may take on the identities of their authors and function as speakers in presentation of communicated discourse. There are 3 examples of prosopopoeia created via IS in the corpus. Consider examples 5.13.a and 5.13.b. In each example the reporting clause (blue) indicates that a work rather than the scientists functions as a speaker. In example 5.13.b "Kaluza-Klein" refers to the Kaluza-Klein theory.

5.13.a Even as far back as the early decades of the twentieth century, a prescient series of papers by the German mathematician Theodor Kaluza and by the Swedish physicist Oskar Klein suggested that there might be dimensions that are proficient at evading detection.

GREENE 2011: 84

5.13.b Kaluza-Klein echoed across the decades answering that the dimensions are all around us but are just too small to be seen.

GREENE 2011: 88

Even though Semino and Short (2004: 106) note that the verbs for the presentation of writing often remain the same as the verbs for speech presentation, I consider examples 5.13.a and 5.13.b presentation of speech and argue that Greene (2011) specially avoided presenting these proposition as writing by choosing to use prosopopoeia. This is one more example of the blurred boundaries between speech and writing in the narratives.

Occurrences of IS like these support Ciapuscio's (2003: 209–210) claim about professional publications being the inspiration for popular texts. They also demonstrate the willingness of the authors to acknowledge the original written text and emphasize its prominence by assigning to it the status of the speaker. Prosopopoeia is the only instance of figurative language that can be clearly attributed to the author of popular science. There is no doubt that it represents the creativity of the narrator and not of the scientists.

The question of attribution becomes less clear, however, when metaphors and analogies figure in Indirect Speech. When the primary function of IS—content presentation—is considered, the conclusion is that IS reflects only the ideas and not the original wording. At the same time, the creative presentation of a scientific issue is a conscious decision and thus could be regarded as part of the content. This line of thought would suggest that the analogies and the metaphors found in IS should be attributed to the scientists. On the other hand, the reformulating properties of IS dictate that the authors are the more likely originators of the figurative language in order to explain the material and make it more relatable. I tend to think that the second proposition is closer to the truth. The major role of non-dramatizing IS with its reformulating as well as summarizing abilities is to present scientific ideas as coming from the scientists yet explained by the authors. (Also see Appendix) Consider example 5.14:

5.14 The authors concluded that we are all immersed in a bath of photons, a cosmic heirloom bequeathed to us by the universe's fiery birth.
GREENE 2011: 39

Popular science authors may capitalize on the freedom to reshape the wording of an utterance that IS provides and choose to restate the original in a simpler and a more engaging way. The example above illustrates Greene's (2011: 39) summary of the result found in "the papers of Gamow, Alpher, and Herman that in the late 1940s announced and explained" the cosmic microwave background radiation—a text which might pose difficulties for a non-specialist. By using IS instead of quoting verbatim, Greene reformulates the utterance as a metaphor, thus making it not only easy for the reader to understand but also creating a sense of the sublime.

Turney (2004a: 91) defines the sublime as "an aesthetic category," with the help of which "Science writers evoke their most telling effects." He goes on to say that "the feeling generated by the sublime includes both awe at the overwhelming sensory impact … and at the human capacity to apprehend it in its full extent" (Turney 2004a: 93). Turney (2004a: 93) also suggests that there could be a more practical application for the metaphors of the sublime since they can function as "the safeguard against the feelings of insignificance induced by cosmic immensities." The fact that the authors choose to incorporate the sublime as part of presented discourse points to a conscious choice to associate the advantages of this strategy (the ability of the human mind to comprehend the universe, anticipation of the possible feelings of insignificance on the part of the reader) with the scientists and thus portray them in a positive light, as creative and caring thinkers.

Turney (2004a: 90) claims that "the sublime has become the characteristic aesthetic of much contemporary popular science." Metaphors of the sublime—the ones similar to example 5.14—have been employed by such celebrity popularizers as Carl Sagan and by the relatively new authors such as Adrian Woolfson (Turney 2004a). Turney (2004a) argues that the phenomenon has been used so often that certain standards have emerged. For example, Turney (2004a: 96) identified "the vocabulary of the sublime" as predominantly seascapes and landscapes. Example 5.15 supports Turney's observations:

5.15 Six years later, in a seminar room at Mount Wilson Observatory in California, Einstein focused intently as Lemaître laid out a more detailed version of his theory that the universe began in a primordial flash and that the galaxies were burning embers floating on a swelling sea of space.

GREENE 2011: 12

Turney (2007: 2) also notes that the use of figurative language is so ubiquitous in popular science that certain metaphors and analogies start to contribute to a pool of stock imagery that many authors "adopt and modify" for their own purposes. In contrast, new metaphors, according to Turney (2004b: 337), indicate that "there is not yet a widely accepted formula for describing … novel" ideas. Turney (2004b: 343) suggests that the adaptation and modification of certain metaphors indicate the success of their originator. There are examples in the corpus that illustrate Turney's (2004 b, 2007) observations. Consider examples 5.16.a and 5.16.b. They share an analogy introduced as IS:

5.16.a … it was, he [Rutherford] remarked, as startling as if a bullet were to bounce off a sheet of tissue paper.

FERRIS 1988: 256

5.16.b It was as if, he [Rutherford] said, he had fired a fifteen-inch shell at a sheet of paper and it rebounded into his lap.

BRYSON 2003: 139–140

To follow Turney's (2004 b, 2007) reasoning, the analogy is repeated in Bryson (2003) because it affords a good way of describing the scientist's reaction. On the other hand, when offering his observations of recycled tropes Turney (2004b, 2007) does not take into account that some of them appear as parts of presented discourse and therefore cannot be automatically attributed to the authors. At the same time, Turney's explanation in this particular case is quite appealing considering the facts that Bryson's text is published after Ferris', that Bryson (2003: 6) acknowledges that he collected his material by "reading books" among other means, and that Bryson (2003: 5) mentions Ferris by name as an example of a science author who writes "the most lucid and thrilling prose." Still, the question of attribution remains. It is unclear whether Ferris (1988: 256) coined this description of the experiment and presented it as coming from the scientist, or whether Rutherford has been recorded expressing his reaction as such an analogy. The absence of references to a source in Ferris' (1988) chapter end notes leaves the question open for the reader and rather suggests that Ferris might have come up with the analogy since, as Toolan (2001: 128) points out, not all Indirect Speech is "necessarily traceable to DD [direct discourse] antecedents." After all, Toolan (2001: 128) notes that "People are quite capable of 'reporting' things that their reportees never said." However, the decision to present the analogy as IS rather than make it a part of narration is significant even if the report of Rutherford's reaction is not accurate. Such view of IS with tropes links it to Tannen's (2007) notion of constructed dialogue and thus to dramatization. However, I am arguing that this kind of IS serves a different primary purpose.

In his exploration of figurative language (particularly metaphors and analogies) in popular science, Turney (2004a, 2004b, 2007) finds that they are used most of the time for explanatory purposes. The fact that the figurative language in presentation of communicated discourse and in presentation of thought occurs in indirect forms (IT and IS) also points to its use for explanation since indirect forms offer the means to reformulate and focus on the content. In effect, the authors are trying to present scientific ideas in more manageable terms for the lay reader and at the same time preserve the place of the scientists as the creators of the content if not necessarily the wording.

Like Ciapuscio (2003), Turney (2004b: 331) explicitly links explanation in popular science with what he calls "re-creation." The types of presented discourse most suitable for this role are the indirect varieties: Indirect Speech, Narrator's Presentation of Speech Acts, Indirect Thought, Narrator's Presentation of Thought Acts. However, as I have shown already (and will continue to demonstrate) NPSA and NPTA are not used for explanations, nor does IT function in this way. This leaves IS as the primary means of presented discourse to introduce explanations. The prominence of figurative language in IS also points to the highly pronounced explanatory properties of this type of presented discourse.

Narrator's Presentation of Speech Acts

It is possible to divide the non-dramatizing NPSA into two groups according to their use of what Gilbert and Mulkay (1984: 57) label "empiricist" and "contingent repertoire." "Contingent repertoire" is an informal scientific discourse that is heavily dependent on "personal or social circumstances," for instance, subjectivity, personal judgment, role of social position, etc. According to Gilbert and Mulkay (1984: 57), "When this [contingent] repertoire is employed, scientists' actions are … depicted … as the activities and judgements of specific individuals acting on the basis of their personal inclinations and particular social positions." As Gilbert and Mulkay (1984: 56) observed, on the other end of the spectrum is the formal, or "empiricist discourse," which creates the image of an objective, "generic" response of a scientist to the phenomena of the natural world.

The majority of the non-dramatizing NPSA (17 out of 29) will fall under the category of empiricist NPSA—they present scientific concepts and ideas. See example 5.17. Contingent NPSA are a slightly smaller group with 12 out of 29 occurrences. They showcase the issues outside of science that nevertheless influence it and present a personal perspective of the scientists on the subject discussed. In example 5.18 it is the reference to the arms race with Russia and the emotional renunciation of the claim that allow me to classify the NPSA as contingent.

5.17 On his sixty-fourth PowerPoint slide, Incandela revealed what you get when you combine these two channels together: 5.0 sigma.

CARROLL 2012: 184

5.18 The American scientific community, which was still grappling with fusion in the fierce race with Russia to produce the H-bomb, declared that the claim was nonsense.

KAKU 2011: 236

Empiricist NPSA do present scientific content, but they hardly explain it. By their nature, they supply a summary which is elaborated on using either narration or other means of communicated discourse presentation, most often Free Indirect Speech. Consider examples 5.19 (NPSA is in blue, narration is in green) and 5.20 (NPSA is in blue, FIS is in red):

5.19 … Italian physicist Dario Autiero announced a result that ended up being more infamous than famous: neutrinos that appeared to be moving faster than the speed of light. The finding came from the OPERA experiment, which tracked neutrinos that were produced at CERN and traveled 450 miles underground to a detector in Italy. Because neutrinos interact so weakly, they can pass through many miles of solid rock with very little loss of intensity, making this kind of arrangement a uniquely effective window onto their properties.

CARROLL 2012: 195–196

5.20 Some argued that Rous had misdiagnosed the tumors; perhaps the injections caused an inflammation peculiar to chickens.

KEAN 2012: 140

Those NPSA that are followed by narration, as in example 5.19, tend to appear at the beginnings of the narratives, with NPSA announcing the subject—usually the discovery discussed. In that, non-dramatizing NPSA resemble IT which presents discoveries. NPSA that are followed by FIS are more likely to be found in the middle of a narrative. In such cases, as example 5.20 shows, FIS adds additional details to the summary provided by NPSA. Example 5.20 also demonstrates how NPSA and FIS can combine to create a hypothesis—again, a function reserved primarily for presentation of thought. Overall, it appears that non-dramatizing NPSA functionally resemble thought presentation.

(Free)Direct Speech

The explanatory function is prominent among the occurrences of (Free)Direct Speech in the corpus. Quite often the authors use (F)DS instead of narration in order to tell the readers how something works or to introduce more details to a concept presented in narration. Consider example 5.21.

5.21 When George Stratton first put reversing goggles on, he had a sense of external motion when he moved his head: "It did not feel as if I were visually ranging over a set of motionless objects, but the whole field of things swept and swung before my eyes."

COEN 2012: 191

Using (F)DS this way the authors once again show the scientists in a positive light—this time, the scientists are able to avoid technical terminology, producing language easily understood by the public, and by doing so, they are also able to incorporate their personal experiences and emotional responses—elements that are likely to engage a lay reader.

The scientists themselves are given the opportunity to explain their experiments and discoveries through (F)DS. See example 5.22.

5.22 ... Stephen Davies of the University of Colorado has had impressive success in treating spinal cord injuries in rats. He says, "I conducted some experiments where we transplanted adult neurons directly into adult central nervous system. Real Frankenstein experiments. To our great surprise, adult neurons were able to send new nerve fibers from one side of the brain to the other in just one week."

KAKU 2011: 128

Note that both occurrences of Direct Speech above contain a degree of dramatization expressed via introduction of personal perspective (example 5.21) and emotionality (example 5.22). The dramatizing properties of these occurrences, however, are less prominent than the explanatory function. These examples show the pervasiveness of dramatization in presented discourse and support my argument that dramatization in various degrees can easily combine with the other functions.

The explanatory and interpretive segments of popular science texts are usually associated with the direct and heavy involvement of the author, which results either in narration or indirect discourse (see, for example, Ciapuscio 2003, Moirand 2003, Turney 2004b, de Oliveira and Pagano 2006). For example, Turney (2004b: 332) regards explanation as "teaching how something is to be understood." The author in this case is seen as the teacher. Even though those who write about explanatory properties of popular science (see, for example, Turney 2004a,b, 2007, Ciapuscio 2003) acknowledge that explanations can be delivered via presented discourse, they do not accentuate such examples. Both Turney (2004b) and Ciapuscio (2003) show that explanations and interpretations originate as direct voices of scientists obtained during private interviews conducted by the authors. However, Ciapuscio (2003: 209) regards explanations and details of the discoveries/experiments supplied by the scientists only as "intermediate stations" of the popularization process, meaning that (F)DS of this kind is not intended as part of a finished popular text but rather serves as a blueprint for the author that he/she will later reformulate and restructure, turning direct discourse into narration. Turney (2004b), on the other hand, suggests that interviews can be presented verbatim as explanations of the scientific concepts.

My observations can confirm that popular science authors use the techniques discussed by both Ciapuscio (2003) and Turney (2004b), incorporating reformulated personal interviews as well as including (F)DS. Example 5.23 shows a lengthy stretch of Direct Speech from a scientist, which Bryson (2003: 208) obtained during a personal interview:

5.23 He [Mike Voorhies] took me to the spot atop a twenty-foot ravine where he had made his find. "It was a dumb place to look for bones," he said happily. "But I wasn't looking for bones. I was thinking of making a geological map of eastern Nebraska at the time, and really just kind of poking around. If I hadn't gone up this ravine or the rains hadn't just washed out that skull, I'd have walked on by and this would never have been found." He indicated a roofed enclosure nearby, which had become the main excavation site. I asked him in what way it was a dumb place to

> hunt for bones. "Well, if you're looking for bones, you really need exposed rock. That's why most paleontology is done in hot, dry places. It's not that there are more bones there. It's just that you have some chance of spotting them. In a setting like this"—he made a sweeping gesture across the vast and unvarying prairie—"you wouldn't know where to begin. There could be really magnificent stuff out there, but there's no surface clues to show you where to start looking."

This example sets DS of the scientist in contrast with IS of the narrator (in blue) who was conducting the interview, thus marking the voice of the scientist as more important. DS takes center stage in this narrative, and narration is used to interject details and set the scene. Voorhies' explanation is clear and is obviously targeting the lay audience of Bryson's (2003) book. Ciapuscio (2003: 209–210) notes that scientists usually are able to "'reformulate' their message according to the addressee" and, in fact, take control "of the content presentation" during interviews, which results in "long monologues." The example from Bryson (2003: 208) is a good illustration of this. However, the text produced by Voorhies is included verbatim as part of a finished popularization and does not function only as an intermediate stage of the writing process—the role it would be more likely to perform according to Ciapuscio (2003).

The fact that the scientists are able to present their findings in a way easily accessible to the public, I suggest, encourages the authors to use (F)DS rather than narration for explanatory purposes. After all, one of the functions of (F)DS in non-fiction is projection of credibility. Using the voices of the scientists to create explanations and supply details of the discoveries enhances the credibility of the authors and their texts while giving the readers an opportunity to experience an unmediated voice of a scientist. Thus it is possible to say that in popular science narratives of discovery the explanatory and the credibility functions of (F)DS converge.

In addition to its explanatory properties, (F)DS can also announce the breakthrough moments in the discovery process—a function that links it with presentation of thought that introduces discoveries and, as I will demonstrate, with narration.

Toolan (2001: 129) suggests that DS may slow down narrative action by focusing on characterization. This is one of the reasons why, as I argued in chapter 4, Narrator's Presentation of Speech Acts are favored for dialogue presentation in the narratives of discovery. At the same time, Semino and Short (2004: 90) point out that "(F)DS is crucial … to the advancement of the plot" in fiction; they do not, however, mention the affects of (F)DS on the speed of narrative pace. The analysis of my data suggests that in the narratives of discovery (F)DS, since it is not the primary means of dialogue presentation, may be used to advance narrative action. In fact, there are several examples where the authors use (F)DS to supply the main portion of the narrative, with the author providing a brief introduction and evaluation of the material. Example 5.24 is one instance of such use of (F)DS, where a major portion of the narrative is presented in the voice of the scientist. Example 5.24 illustrates how (F)DS can be used to showcase the eureka moment of the discovery:

5.24 The idea of natural selection first occurred to Darwin shortly after he returned from his five year odyssey on the *HMS Beagle* ... As a result of the trip, Darwin became convinced that species were not fixed but could be modified over time. But he did not know of a mechanism that could explain how species change and adapt. Then ... he was reading a book on population growth by Thomas Malthus. Malthus had pointed out that ... the population size would eventually outstrip the food supply, and struggle and starvation would ... follow. This idea struck home:

> I happened to read for amusement Malthus on *Populations*, and being well prepared to appreciate the struggle for existence which everywhere goes on from long-continued observation of the habits of animals and plants, it at once struck me that under these circumstances favourable variations would tend to be preserved and unfavourable ones destroyed. The result of this would be the formation of new species. Here, then, I had at last got a theory by which to work.

COEN 2012: 24–25

Coen (2012) could have easily presented this pivotal moment in Darwin's career using narration. However, using FDS shows Darwin himself as a capable storyteller, able to supply the information using language suitable for a lay reader.

In instances like this, the author sets up the narrative by supplying the reader with the relevant background information (blue portion of example 5.24, which includes indirect forms of discourse presentation) but then disengages from the story, leaving the scientist to move the narrative forward from that point on. Thus FDS takes on the functions of narration. This is demonstrated even more clearly in the case of a narrative from Carroll (2012: 68–69) that embeds DS and NPSA inside (F)DS that is used as narration. Example 5.25 shows the portion of the narrative related via (F)DS with embedded DS (green) and embedded NPSA (blue):

5.25 Completed in 1983, the Tevatron was the highest-energy accelerator in the world until LHC [Large Hadron Collider] took the crown in 2009. Its crowning achievement was the discovery of the unusually massive top quark … in 1995. Gordon Watts … remembers the moment when the signal climbed above the important "three sigma" threshold [an indication that a new particle is discovered]:

We were in one of the big top meetings reviewing all the analyses that were about to go out for one of the conferences. Every analysis was seeing a small excess, but it was so small that it wasn't really meaningful … one of the people … raised his hand … "Uh … hold it a moment … if I do the simplest thing here and add up all the backgrounds and the signals I get over three sigma." There was a silence in the room while everyone went scrambling back through the talks to figure out if that was actually correct. Either the spokesperson or the top convener spoke next … it was a four-letter word. I think everyone felt the chill go down their spine.

Semino and Short (2004: 171–175) also noted that FDS can host embeddings. However, the examples they supply are relatively short and by no means represent whole stories. Thus, it would be hard to make an argument that FDS in their corpus takes on the functions of narration. Free Direct Speech in the popular science narratives of discovery when presenting the whole stories resembles first-person narration in fiction when the narrator is also one of the characters in the story. In such situations, the voice of that particular character-narrator seizes to be an instance of discourse presentation and is regarded as pure narration. I suggest that in the narratives of discovery lengthy occurrences of FDS function the same way. Whenever a scientist engages in a monologue that contributes a significant portion of a narrative, he/she becomes the narrator for the duration of FDS. Such use of FDS is an example of fictionalizing the narratives.

The willingness to relinquish the telling of the story to the scientists points to a kind of respect and solidarity between the authors and the scientific community, which is also evident in the near absence of the distancing function of (F)DS. The authors and the scientists present a united front when it comes to addressing scientific issues. The only occurrence of the distancing function of Direct Speech is shown in example 5.26.

5.26 In response to the question "If string theory is the long-sought unified theory, then why haven't we seen the extra dimensions it needs?" Kaluza-Klein echoed across the decades answering that the dimensions are all around us but are just too small to be seen.

GREENE 2011: 88

Direct Speech in this case is unattributed, which indicates a hypothetical and a general disagreement rather than a concrete objection. By placing the question in quotation marks, Greene is acknowledging the presence of the issue without being responsible for raising it. The positive response to the question presented as Indirect Speech (blue) contrasts with the DS used to pose it. By choosing a more author-controlled mode of speech presentation to give the answer, Greene demonstrates his support for the latter.

The near absence of the distancing function of (F)DS in the corpus could be explained by the overall purpose of popular science—that is, to promote scientific advancements and to connect the public with the scientists. Unlike journalistic writing, which even when describing scientific subjects, tends to dwell on social controversy (see, for example, Calsamiglia and Ferrero 2003, Moirand 2003), popular science books are more concerned with knowledge transfer and with presentation of science as a positive force. Distancing themselves from the voices of the scientists will not help these authors achieve their goals. Expressing solidarity with presented discourse is much more productive. Myers (1999: 389) notes a similar use of presented discourse in oral group discussions, where the speakers express solidarity with the presented utterances. At the same time, analyses of (F)DS in written non-fiction (see, for example, Bell 1991, Calsamiglia and Ferrero 2003, Moirand 2003, Semino and Short 2004, de Oliveira and Pagano 2006, Smirnova 2009, Urbanova 2012) tend to focus on the distancing function and do not mention that (F)DS can be employed "to signal solidarity" as well as "suggest detachment between the reporting and reported utterances" (Myers 1999: 389).

Fu and Hyland (2014: 123) argue that popular science is "persuasive, seeking to convince the reader both of the importance of the content and a wider ideology of scientific progress." This goal would be hard to attain if the authors distanced themselves from the practitioners of science. Instead, the authors choose to praise the scientists and their achievements. Bucchi (1998: 14) calls this kind of popularization technique "'celebratory' discourse." While Bucchi (1998) explores the multiple linguistic strategies that can help the authors create such a discourse, he does not distinguish between narration and presented discourse, and, in fact, seems to focus on narration alone. My observations indicate that presented discourse of scientists is a significant contributing factor to celebratory exposition of science.

Conclusion

It is commonly assumed (see, for example, Semino and Short 2004, Leech and Short 1981/2007, Short 2007, 2012) that presentation of communicated discourse and presentation of thought have radically different functions. At the same time, my exploration of presented discourse shows that in the narratives of discovery there are many functions that presentation of communicated discourse and presentation of thought share. The greatest difference between the two appears to be the degree of dramatization. Unlike other studies (see, for example, Toolan 2001, Semino and Short 2004), however, my analysis suggests that presentation of thought in the narratives of discovery has very little to do with dramatization. Instead of revealing the inner worlds of the scientists (and thus dramatizing the narratives), presentation of thought is more closely connected with the scientific matters than it is with the intimate thought processes. In fact, the presence of the verb patterns associated with the presentation of hypotheses and discoveries via thoughts of scientists functions as a mitigating strategy that suggests the general rather than the individual nature of the information introduced.

By bringing the scientific rather than intimate issues to the forefront, presentation of thought is functionally connected to non-dramatizing communicated discourse, which too is science-focused and most often is devoted to the explanation of scientific concepts. Overall, presentation of thought in the corpus is almost devoid of emotions and does not possess any of the emotional markers associated with dramatizing communicated discourse. This does distinguish presentation of communicated discourse from presentation of thought but not for the reason most often highlighted in other studies where presentation of thought is seen as the primary carrier of dramatization.

A function that firmly connects presentation of thought with non-dramatizing communicated discourse is celebration of science and scientists. In fact, these two modes of discourse presentation could be united under Bucchi's (1998) label "celebratory discourse." Bucchi (1998: 14) understands celebratory discourse as an "outcome of communication at the public level" that "further strengthens the certainty and solidity of theories and results." My analysis demonstrates that presented discourse expands the reach of the celebratory discourse to include not only science in general ("theories and results") but also individual scientists. As presentation of thought indicates, the scientists are much more likely to come up with positive hypotheses than with ideas that find no empirical ground. Both presentation of thought and non-dramatizing communicated discourse show scientists as creative thinkers who are capable of producing engaging and meaningful explanations of their findings. Presentation of communicated discourse also reveals the story-telling talents of the scientists. Overall, presentation of thought and non-dramatizing communicated discourse introduce a portrait of an idealized scientist who is not only accomplished on a professional level but is also willing and able to communicate his/her discoveries to the curious public.

The implications of this almost exclusively positive portrayal of scientists and, by extension, science as an activity are addressed more fully in chapter 7.

CHAPTER 6

Presented Discourse outside the Narratives of Discovery

Introduction

Until now, I have focused on presented discourse in one specific part of popular science books—narratives of discovery. The voices and thoughts of scientists, of course, could be found throughout the books. The reason for looking at narratives was to determine how presented discourse enhances explanatory segments. Now the time has come to broaden the scope of the analysis and consider the functions of presented discourse outside the narratives of discovery. In this chapter, I provide an overview of presented discourse in popular science books and take a second look at the phenomena observed inside the narratives, this time in a broader context. The goal is to create a fuller picture of the role presented discourse plays in popular science and to test the claims about the nature of presented discourse of scientists made on the basis of the analysis of the narratives of discovery. The comparison between the functions of presented discourse inside and outside the narratives reveals several important points.

Firstly, the narratives present the scientists' discourse as science-centered (this is especially evident in presentation of thought) and produce a picture of scientists who are predominantly research-oriented. In contrast, presented discourse outside the narratives contributes toward a more balanced portrayal, with the scientists shown as occupied by personal as well as professional issues.

Secondly, presented discourse outside the narratives uncovers a variety of voices, not just those of the scientists, while the narratives contain presented discourse of scientists almost exclusively. One of the new voices that I find particularly interesting is connected with the issue of fictionality; it is the voice of the reader, who through discourse presentation effectively becomes a character in the books. I propose that introducing presented discourse of the reader goes beyond the construal of the "reader-in-the-text" that Thompson and Thetela (1995) and Thompson (2001, 2012) discuss. What I shall call the "fictionalized reader" is a more salient and concrete presence than Thompson's (2012: 80) "virtual entity, which haunts all discourse." The fictionalized reader is an embodied presentation with a clear voice.

© KONINKLIJKE BRILL NV, LEIDEN, 2018 | DOI: 10.1163/9789004365971_007

The construction of the fictionalized reader is not the only fictionalization technique evident in presented discourse outside the narratives. In chapter 4, I mentioned that dialogue incorporated via (Free)Direct Speech is nonexistent in the narratives of discovery. This popular feature of fiction (according to Semino and Short 2004: 90, for example) is, however, part of presented discourse outside the narratives and could be seen as a fictionalization mechanism. The presence of the fiction-like elements outside the narratives points to the overall tendency for fictionalization in popular science.

Thirdly, a look at presented discourse outside of the narratives uncovers an interdisciplinary side of popular science. It indicates the strong connections that the authors draw between science and literature. References to literary works as well as presentation of communicated discourse and thoughts of writers are used as support for the points the authors make.

The presentation of discourse of other writers is directly connected with presentation of writing, which I have chosen to consolidate with the speech presentation in the analysis of the narratives because of the difficulties in distinguishing between speech and writing. In presented discourse outside the narratives, however, the line between speech and writing is significantly less blurred, allowing me to confirm Semino and Short's (2004) proposition that writing presentation should be regarded as a separate category of presented discourse alongside speech and thought presentation. Outside the narratives of discovery, it becomes clear that the presentation of writing is a viable category of presented discourse in popular science.

The sections that follow elaborate on each of the aspects introduced above and offer examples and comparisons. This chapter provides the additional context and background against which presented discourse in the narratives of discovery can be evaluated. The analysis presented here combined with the data from the previous chapters also allows for certain generalizations of the findings concerning presented discourse in popular science.

Celebratory Discourse

I concluded the previous chapter suggesting that presented discourse in the narratives of discovery, and especially presentation of thought and non-dramatizing communicated discourse, produce a celebratory presentation of science and scientists. I identified specific types and functions of presented discourse that are used to achieve this goal in the narratives—namely Indirect Thought and Narrator's Presentation of Thought Acts that introduce positive hypotheses and Indirect Speech that shows scientists as creative thinkers. In this section, I will further explore the resources of presented discourse devoted to promoting a positive outlook on science.

"A Neutrino Walks into a Bar": Humor in Presented Discourse

Presented discourse of scientists in the narratives of discovery tends to cover science-related topics exclusively. This is especially evident in the analysis of thought presentation, which fails to reveal intimate thoughts and observations in favor of projecting science-centered mental processes. As I mentioned in chapter 5, such treatment of presented discourse creates a picture of scientists who are dedicated to their work but fails to reveal their personalities. Presented discourse outside the narratives remedies this situation by including personal thoughts and even jokes from scientists introduced in their own voices. For example, Bryson (2003: 30) uses Direct Speech to juxtapose Bob Evans' extraordinary talent for spotting supernovae with his lack of other, more practical, skills, thus showing the scientist in a more personal light. Consider example 6.1.

6.1 "I just seem to have a knack for memorizing star fields," he told me … "I'm not particularly good at other things," he added. "I don't remember names well."

"Or where he's put things," called Elaine [Evans' wife] from the kitchen.

This example demonstrates another characteristic of presented discourse absent in the narratives of discovery—the voices of the non-scientists. These voices that include spouses, biographers, and colleagues often supply evaluations of the scientists, taking them out of the realm of science and into the realm of the personal. This function is usually associated with presentation of thought (see chapter 5), but in popular science, it appears, it can be fulfilled by communicated discourse. Indirect forms of communicated discourse alongside (Free)Direct Speech can also showcase the personal sides of the scientists. Example 6.2 uses Narrator's Presentation of Speech Acts (in blue) to display Buckland's dietary preferences:

6.2 Depending on whim and availability, guests to Buckland's house might be served baked guinea pig, mice in batter, roasted hedgehog, or boiled South Asian sea slug. Buckland was able to find merit in them all, except the common garden mole, which he declared disgusting.

BRYSON 2003: 69

In addition to displaying skills and preferences of the scientists outside the laboratory, presentation of communicated discourse can also be used to incorporate humor originating with the scientists into the texts. Examples 6.1 and 6.2 may be considered somewhat amusing, but the use of NPSA instead of (F)DS in example 6.2 points to the author as the source of humorous presentation. In example 6.2, the reader encounters essentially Bryson's (2003) take on Buckland's eating habits; there is no indication that Buckland himself regarded his unusual menus with anything but scientific curiosity. When (F)DS is used to introduce humor, it is unmistakably associated with the speaking scientist thus showing him/her as capable of a lighthearted approach to scientific issues. Consider example 6.3.

6.3 Another physicist at the party, Gordon Watts of the University of Washington, was asked whether the long anticipation for the LHC has been successful.

"Oh yeah, completely. I have this shock of gray hair here now. My wife claims it's because of our kid, but it's really because of LHC."

CARROLL 2012: 2

Besides introducing humor via DS, example 6.3 also demonstrates several other characteristic properties of presented discourse outside the narratives of discovery. There is an interaction of discourse presentation types that represents a dialogue: NPSA (in blue) introduces a question from an anonymous speaker, which receives a not-altogether serious response via DS of the scientist, which in turn contains embedded IS of his wife (in green). The juxtaposition of the wife's IS with the DS of the scientist creates the humorous effect. Secondly, as I suggested above (see example 6.1) the introduction of voices that do not belong to the scientists tends to shift the focus of presented discourse from science to the more intimate setting of home and to personal responsibilities. In example 6.3, the wife's voice reveals the scientist as a father, while his own DS puts the focus back on his work—the Large Hadron Collider.

Other instances of presented discourse, however, show scientists venturing outside of their professional sphere (usually to find analogies and draw comparisons) without any help from other speakers. Kean (2012: 151) provides a good example (presented discourse of the scientist is in blue):

6.4 ... a lab-tweaked version of one virus can turn polygamous male voles—rodents who normally have, as one scientist put it, a "country song ... love 'em and leave 'em" attitude towards vole women—into utterly faithful stay-at-home husbands ...

Showing scientists as having a humorous and lighthearted attitude toward their research and as being able to incorporate popular cultural references may create a more balanced picture when this portrayal is considered against the one supplied by the narratives. Outside the narratives of discovery, scientists emerge as having one more dimension to their personalities. I regard this as a contribution to celebratory discourse. Through the use of humor in presented discourse, the authors produce a picture of science and scientists that is rooted, in some way, in a universally positive emotion.

"All Healthy Bodies Resemble Each Other, While Each Unhealthy Body is Unhealthy in Its Own Way": Literary References in Presented Discourse

Another aspect of presented discourse outside the narratives of discovery that I consider significant to the celebration of science is the connection between science and literature. It can manifest as DS in the form of a quote from a famous literary work or through IS or IT of well-known writers. Examples 6.5.a, 6.5.b (IS is in green), and 6.5.c (IT is in blue) illustrate these occurrences:

6.5.a ... William Shakespeare was born ... in a plague year ..., and his career was interrupted several times, when plague epidemics forced the theatres to close down. Shakespeare had Mercutio, in *Romeo and Juliet*, say, "A plague on both your houses!" ... His audiences would have understood what he meant. Most doctors thought that plague was a new disease, or at least one that Galen had not written about ...

BYNUM 2012: 42

6.5.b As well as responding to individual events, animal nervous systems can also respond to sequences of events over time. If you repeatedly stimulate the slug's siphon, the gill-withdrawal reflex progressively weakens ... It is a case of what Marcel Proust called the anaesthetizing effect of habit ...

COEN 2012: 141

6.5.c Verne, more than others, realized that science was the engine shaking the foundations of civilization ...

KAKU 2011: 5

The inclusion of presented discourse of literary figures and their characters establishes science as a discipline that does not stand alone nor is in opposition to other forms of human inquiry but functions in accord with them. It is worth noting that presented discourse connected with literature works as a supporting mechanism for the scientific issues discussed, thus evaluating the subjects at hand as recognizable and valuable outside the circle of professional scientists.

Presented discourse of this kind works as a reader engagement mechanism, as it alludes to the reader's possible familiarity with certain literary figures and their works and uses those to make the new, scientific information more relatable. The degree of assumed familiarity of the reader with a certain literary work, of course, varies from text to text, and some authors include introductions of writers whose works are presented (as in example 6.5.a), and some rely on the reader's ability to recognize the writer and the work without any help. This last method allows the popular science authors to present fictional literary discourse as a source of wordplay and jokes. Example 6.6 shows Kean's (2012: 300–301) use of Tolstoy's opening lines in *Anna Karenina* to illustrate a process of chromosome sequencing:

6.6 As observers have noted, the process was analogous to dividing a novel into chapters, then each chapter into sentences. They'd photocopy each sentence and shotgun all the copies into random phrases—"Happy families are all," "are all alike; every unhappy," "every unhappy family is unhappy," and "unhappy in its own way."

The first part of the title for this subsection is another example of Kean's (2012: 312) use of the same novel as a source of wordplay attributed to the scientists. Example 6.7 provides the full sentence from Kean in context:

6.7 … even if the large-scale symptoms are identical, the underlying genetic causes … might be different. (Some scientists misquote Tolstoy to make this point: perhaps all healthy bodies resemble each other, while each unhealthy body is unhealthy in its own way.)

This incorporation of a distorted quotation into presented discourse is similar to the use of figurative language in indirect discourse (see chapter 5). In both cases the creative use of language is attributed to the scientists, showing them as versed in subjects other than the ones required for their professions.

Including humor and literary references into presented discourse, and especially into presented discourse of scientists, positions scientific endeavors as part of universal human knowledge and experience. Presented discourse that shows science as applicable and relatable to a variety of human activities contributes to celebrating its achievements.

Speech Presentation versus Writing Presentation: Another Look

In chapter 3, I argued for not distinguishing between speech and writing presentation in the corpus of the narratives of discovery because unambiguous references to writing were few and a clear distinction between speech and writing was not always possible. This is why I have been using the term communicated discourse, which includes both speech and writing presentation.

The investigation of presented discourse outside the narratives of discovery, however, demonstrates that the amalgamation of the two discourse presentation categories is useful in the analysis of the narratives exclusively and does not need to be preserved when working with other parts of the texts. Presentation of discourse outside the narratives of discovery makes the distinction between speech and writing much clearer, and the references to writing are not only numerous, but would also lose part of their significance if analyzed as speech presentation—which is not the case inside the narratives of discovery. References to literature and to specific works of literary fiction are in some measure, I think, responsible for the clearer lines between writing and speech. In such cases as in examples 6.5.a, 6.6, and 6.7, presented discourse introduces quotations from literary works, not the discourse of individual authors. For example, when Bynum (2012: 42) quotes Shakespeare, it is important to let the reader know that he is quoting the play *Romeo and Juliet* and not something Shakespeare possibly said or thought. In other words, presented discourse introduced as presentation of writing reflects Shakespeare the author, who expresses himself through his characters, as opposed to Shakespeare the person, who would speak/write as himself (in a letter to a friend, for example, if such a thing existed). The same is true for Kean's (2012: 300–301, 312) examples (6.6 and 6.7), as they deal with Tolstoy the author, who is projected through the voice of the narrator in *Anna Karenina*.

Furthermore, writing presentation is distinguishable from speech in occurrences other than the ones dealing with literary works. For example, Ferris (1988: 331–332) refers to Howard Georgi "writing a limerick on the blackboard." Carroll (2012: 78) quotes a caption that appeared under "a photograph of the CMS detector next to a photograph of a pigeon" in the *Telegraph*. See examples 6.8.a and 6.8.b.

6.8.a ... Howard Georgi, known for his work in grand unified theory, introduced a 1984 Weinberg lecture at Harvard by writing a limerick on the blackboard that read:

Steve Weinberg, returning from Texas,
Brings dimensions galore to perplex us.

But the extra ones all
Are rolled up in a ball
So tiny it never affects us.

6.8.b The *Telegraph* printed a photograph of the CMS detector next to a photograph of a pigeon, with the caption, "The Large Hadron Collider (left) and its arch-nemesis (right)."

These and similar occurrences are unmistakably presentations of writing and explicitly intended as such. Introducing them indicates that the authors, on these particular occasions, wish for the reader to process the information as having come from a written source. Mistaking writing for speech would either alter the facts (as in the case of Georgi) or would make the discourse nonsensical (in the case of the *Telegraph*'s caption). Presented writing is usually associated with additional artifacts such as specific forms of writing (e.g. limericks, novels, etc.) or images (as in example 6.8.b). Once divorced from these scaffolds, the presentation of writing may become unclear to the reader. In a way, what this evidence shows is that writing is a more constrained form of presented discourse as compared with speech. When the scaffolding background is less important, writing can be substituted for speech—as is the case in the narratives of discovery. However, once the background scaffold is vital to the correct interpretation of the message, the presentation of speech cannot be substituted for writing.

This idea is supported in the findings of Short et al. (2002) and Semino and Short (2004: 113), who suppose the presentation of writing more accurate and attentive to the details of the anterior discourse than the presentation of speech. The suggestions of Short et al. (2002) are of particular relevance. Short et al. (2002: 327) argue for a "context-sensitive account of discourse presentation" that gives writing presentation the monopoly on accuracy because of the "checkability" factor. This attention to context and the potential for checking an instance of writing presentation for accuracy is connected with what I have been referring to as "artifacts" and "scaffolding background" of presented writing.

One possible explanation for more explicit references to writing outside the narratives of discovery is the general broader range of the topics covered by presented discourse and also the broader range of the presented voices, some of which are available only through writing and become less effective as presented speech, as is the case with some literary references. Another possible reason is that outside the narratives of discovery the broader focus allows for introducing contexts other than those of communication between scientists.

Examination of presented discourse outside the narratives of discovery confirms the need for the introduction of a separate category of writing presentation as suggested by Semino and Short (2004).

Confirming the Fusion of Non-fiction and Fiction-Like Qualities of Presented Discourse in Popular Science

In general, it is possible to say that presented discourse outside the narratives of discovery is functionally similar to presented discourse in the narratives. It too is used to dramatize and to present factual information and thus demonstrates a fusion between the functions usually associated with fiction and non-fiction. However, presented discourse outside the narratives appears to preserve the distinction between fiction and non-fiction when it comes to the form presented discourse takes and the functions it performs, while the narratives blend the forms and functions more thoroughly. Comparison of (Free)Direct Speech outside and inside the narratives of discovery provides some of the most vivid examples of the formal contrasts and functional similarities.

As Semino and Short (2004: 93) demonstrate, (F)DS in non-fiction (especially in newspapers) is mostly used to incorporate facts and, when necessary, distance the author from the presented voice. I showed in chapter 4 that a significant portion of (F)DS in the narratives of discovery is used for dramatization—a function reserved primarily for (F)DS in fiction (see, for example, Toolan 2001, Semino and Short 2004). In chapter 5, having analyzed the non-dramatizing occurrences of (F)DS, I came to the conclusion that (F)DS can be used for explanatory purposes and can even contribute significantly to narrative progression. Semino and Short (2004: 90, 93) note that (F)DS in non-fiction manifests as "individual utterances in isolation," while in fiction it is more likely to create "stretches of dialogue." In fact, I demonstrated in chapter 4 that (F)DS in the narratives of discovery is not used to create dialogue and that this role is performed by Narrator's Presentation of Speech Acts instead.

Outside the narratives of discovery, however, there are numerous examples of fiction-like dialogic exchanges presented via (F)DS, which align popular science more closely with fiction than with non-fiction. Consider example 6.9.

6.9 "Would you do me the favor," said Frederick to the darkling Lambert, "of telling me in what sciences you are specialized?"

"In all of them," Lambert replied, addressing a point in space ninety degrees away from the king.

"Are you also a skillful mathematician?" asked Frederick.

"Yes."
"Which professor taught you mathematics?"
"I myself."
"Are you therefore another Pascal?" asked Frederick, referring to the great mathematician of the previous century.
"Yes, Your Majesty," replied the voice in the dark.

FERRIS 1988: 149

At the same time, presented discourse outside the narratives also contains short, factual, isolated utterances that Bell (1991) and Semino and Short (2004) consider prototypical for non-fiction. Consider example 6.10.

6.10 Commenting on the report, the editors of *Morbidity and Mortality Weekly Report* speculated that the puzzling symptoms of the five men "suggest the possibility of a cellular-immune dysfunction."

ZIMMER 2011: 56

As examples 6.9 and 6.10 show, (Free)Direct Speech outside the narratives of discovery conforms to functions generally associated with both fiction and non-fiction. Presented discourse in the narratives of discovery acts similarly, but in the case of (F)DS the form and the function do not always correspond as belonging to fiction or non-fiction as they do outside the narratives. For example, shorter stretches of isolated Direct Speech are more likely to dramatize by supplying personal perspective and emotionality (form associated with non-fiction, function associated with fiction)—see example 6.11a. At the same time, longer occurrences tend to incorporate more factual information (form associated with fiction, function with non-fiction)—see example 6.11.b.

6.11.a Even at that time it was becoming clear that the basic outline of the Standard Model was "pretty much a done deal," as he [Wilczek] put it, but that the properties of the Higgs boson were relatively unexplored.

CARROLL 2012: 169

6.11.b Pavlov noticed that many factors could trigger the dog's salivation, as long as they were associated with presentation of food: "Even the vessel from which the food has been given is sufficient to evoke an alimentary reflex complete in all its details; and, further, the secretion may be provoked even by the sight of the person who brought the vessel, or by the sound of his footsteps."

COEN 2012: 153

As Semino and Short (2004: 90–93) observed, for (F)DS in fiction, longer stretches tend to dramatize, for (F)DS in non-fiction, short quotes are more likely to be devoted to the presentation of facts. By changing the alignment between the form and function of (F)DS the narratives of discovery show a greater integration of both fiction and non-fiction-like qualities of presented discourse, while presented discourse outside the narratives remains more identifiable as resembling either fiction or non-fiction.

A feature that aligns presented discourse outside the narratives with non-fiction is the distancing function of presented discourse. This property is usually associated with DS (see, for example, Semino and Short 2004: 93). It should be remembered that DS with distancing properties is absent in the narratives except for one undisputed occurrence introduced in chapter 5. (See chapters 4 and 5 for a discussion of distancing in DS as it relates to the narratives of discovery.) Outside the narratives, however, it is easy to find examples of DS used to create the distancing effect. Consider example 6.12.

6.12 In 1983, as part of the World's Columbian Exposition in Chicago, seventy-four well-known individuals were asked to predict what life would be like in the next 100 years ... Senator John J. Ingalls said, "It will be as common for the citizen to call for his dirigible balloon as it now is for his buggy or his boots."

KAKU 2011: 7

In this case, and in all the other cases of distancing in the books I analyzed, the author distances himself from what is currently regarded (usually by him) as incorrect information. For example, Greene (2012: 9) explicitly shows his non-alignment with the presented voices because they do not express the point of view he favors:

6.13 Some people recoil at the notion of parallel worlds. As they see it, if we are part of a multiverse, our place and importance in the cosmos are marginalized. My take is different.

Here indirect forms of presented discourse (Narrator's Presentation of Thought Acts is in blue and Indirect Speech is in green) are used to distance the author from the views introduced.

Having seen that outside the narratives of discovery other forms of presented discourse than DS can be used to distance the author from the material, it is possible to regard the negative hypotheses presented via IT and NPTA as instances of distancing. Negative hypotheses introduce information that does not find empirical support and thus is considered incorrect by the scientific community (see chapter 5). The verb choices associated with the presentation of the negative hypotheses can be seen as signals of distancing.

Distancing, dramatization, and presentation of facts are all confirmed functions of presented discourse inside and outside of narratives and represent a blend of non-fiction and fiction-like qualities of popular science. The brief comparison of these functions shows that in the narratives of discovery the functions associated with non-fiction and with fiction are fused together and appear more difficult to discern because the forms of presented discourse and their respective prototypical functions do not always align. For instance, Direct Speech can be used to introduce explanations and Narrator's Presentation of Speech Acts to create dialogue, while Indirect Thought and Narrator's Presentation of Thought Acts can produce the distancing effect, and Indirect Speech can introduce figurative language showcasing not content but form; not to mention the preference of thought presentation for scientific rather than personal matters. Outside the narratives, both non-fiction and fiction-like properties of presented discourse are still present, but they are usually aligned with the prototypical forms. Thus DS can be used to create dialogue (fiction-like quality) and for distancing purposes (non-fiction-like quality). Both of these functions are clearly identifiable as prototypical for DS in fiction and non-fiction.

I suggest that such a division of functions as observed outside the narratives of discovery can be attributed to a generally lesser need to use presented discourse for dramatization. When dramatization using presented discourse outside the narratives does occur, it takes on a form that is very different from anything in the narratives of discovery.

The Fictionalized Reader in Popular Science

The notion of the reader as reflected in a text is a well-documented phenomenon (see, for example, Iser 1972 and Culler 1982, who account for such a reader in fiction, or Fairclough 1989 and Talbot 1995, who look at non-fiction). More recent examples include Hyland (2001, 2005, 2009, 2010), who explored the reader as it manifested in academic writing, and Perrault (2008), who looked at popular science. The idea of the reader as projected by the author has become known as the "reader-in-the-text"—a label first proposed by Thompson and Thetela in 1995 and later explicated in more detail by Thompson (2001, 2012).

Despite the wealth of studies on the subject and the different approaches to the reader-in-the-text (see, for example, Martin and White 2005 or Lewin and

Perpignan 2012), one basic underlying principle remains. That is the assumption that the reader-in-the-text, even though a construct of the author, is still essentially a reflection of a real-life reader whose needs the author should anticipate and address. The kind of reader-in-the-text that emerges from this view of the phenomenon is at best an approximation of a generalized audience. It is an ethereal entity that, in Thompson's (2012: 80) own words "haunts all discourse." The presence of the reader-in-the-text is usually revealed through analysis of interactional metadiscourse (Kylinich and Pokalyhina 2013, Fu and Hyland 2014) or evaluative techniques employed by the author (see, for example, Hunston 2000, Hunston and Thompson 2000). Such analyses are designed to bring to the forefront the covert mechanisms by which the reader manifests. The notion that the reader is always presented covertly has not been questioned so far, nor have there been suggested any alternative ways of incorporating a reader in a non-fiction text. Thompson (2001: 59–60) does discuss "assigning speech roles to [authors] and the readers," but his examples indicate that he is still dealing with the reader as being part of narration and therefore being presented in the voice of the author. Below is an example from Thompson (2012: 79) which shows how a typical reader-in-the-text may appear:

6.14 But I really shouldn't care about that stuff, he's a great guy. But really … socks with sandals … all the time?

Thompson (2012: 79) explains, that "the reader of the blog is projected as sharing with the writer the culturally-specific belief that for men to wear socks with sandals is a sartorial blunder." While the author gets to ask a question, the reader-in-the-text does not get to respond openly. Thompson (2001: 60) explains the covert and voiceless reader-in-the-text by observing that "in a written text it is clearly not a straightforward matter to include a response from the reader."

This is true if the author follows the above-stated assumption that the-reader-in-the-text is a representative of a multitude of real readers. It is impossible to guess and include responses from all the potential readers, so the reader-in-the-text is resigned to silence. According to Thompson (2001: 60; 2012: 80), there is also no guarantee that a possible response that a text constructs in such a situation (as it inevitably must in order to proceed) will correspond to how an actual reader will react. The assumption that the reader-in-the-text is a real yet an indeterminate entity creates a real problem for writers of argumentative and explanatory texts which rely heavily on interactional mechanisms. The writer is forced into a perpetual web of guesswork that is not guaranteed to produce the reactions he/she expects from his readers.

The traditional reader-in-the-text has been explored most fully in argumentative texts (see, for example, Thompson 2001, 2012; Hyland 2005). The inability to identify precisely the position the reader occupies in such texts might not be as inconvenient as it is in explanatory texts. For instance, Thompson (2001: 60) shows that when a response from the reader is necessary for the progression of the argument, the writer acts as if an answer satisfactory to his/her position has been supplied. The lack of voice on the part of the reader-in-the-text becomes advantageous since it allows the writer to proceed without the distraction of an unfitting response.

When it comes to explanatory texts, the situation is different. Interactional mechanisms in such texts are often used not so much to advance the agenda of the author but to check whether or not the reader understands the information presented. Thus the requirements placed on the reader-in-the-text are different. Popular science, an essentially explanatory genre that relies on interactional discourse to relate complex scientific issues to non-specialists, creates an alternative version of a reader-in-the-text. This is now an entity that does not always have to be a covert, voiceless shadow of a real-world reader but may become an entirely fictional yet much more concrete and openly presented personage. Popular science writers have created a new kind of a reader-in-the-text, which I shall label "the fictionalized reader." The most salient manifestation of the fictionalized reader is through presented discourse. The fictionalized reader is a reaction by the popular science writers to the indeterminacy of the traditional reader-in-the-text, which undermines explanatory mechanisms available to a popular science author.

The familiar mechanisms for including the reader in a text include evaluation, modalization, reader pronouns, and others (Thompson 2012: 81, Kylinich and Pokalyhina 2013: 767) and deal with covert representation of the reader. The fictionalized reader in popular science, on the other hand, manifests through its own presented discourse and functions like any other presented voice in a text. That is, the fictionalized reader is treated by the popular science author (and is to be perceived by the reader) the same way a scientist is, as an actant in the text.

As I demonstrated in the previous chapters, analysis of presented discourse in the popular science books reveals that the authors employ a variety of presented discourse types to showcase the scientists whose achievements they discuss. In addition to using presented discourse to relate scientific information as coming to the reader from the scientists themselves, the authors also rely on presented discourse to display what Gilbert and Mulkay (1984) identified as "contingent discourse." That is presented discourse of scientists that adds personal perspectives and incorporates issues that lie beyond the realm of science. In general terms, the use of presented discourse brings to popular science dramatization and emotionality thus ultimately presenting scientists as relatable and accessible.

The fact that the reader is also assigned his/her own presented discourse points to an attempt to integrate the reader into the texts more fully and to create a forum where a non-specialist can take part in scientific experiments. According to Votrina (2012) and Kylinich and Pokalyhina (2013), any kind of presented discourse in popular science books is a signal of dialogicity. The actual involvement of the reader with the text through specially created presented discourse of the reader shows the authors who use this technique as challenging a "canonical" or a "dominant view" of science popularization.

A canonical or a dominant view is a popularization model that acknowledges only a one-way communication process: from the enlightened experts to the ignorant public. Myers (2003: 266) outlines "several assumptions that go with this view." The fictionalized reader directly contrasts with two of them: the idea that scientific "knowledge travels only one way, from science to society" and the notion "that the public sphere is, on scientific topics, a black slate of ignorance on which scientists write knowledge" (Myers, 2003: 266).

In recent years, it has been shown that the public does contribute to the scientific community (Moirand 2000, 2003; Luzón 2013). At the same time, the kind of contribution offered by the modern layman is radically different from the way non-specialists engaged with science in the past. If in the 18th and 19th centuries a lay person could actually produce scientific knowledge (Lightman 2000, Topham 2000, Bensaude-Vincent 2001, Oparina 2015), in the 21st century, he/she is more likely to evaluate it, engage with it on an emotional level, in other words, take part in it indirectly (Calsamiglia 2003, Moirand 2003, Myers 2010, Laslo et al. 2011, Luzón 2013, Supper 2014). The fictionalized reader is one manifestation of the lay public taking part in science. Recent research that deals with interaction of experts and non-experts in popular science texts demonstrates that the authors (who are often scientists) are treating the readers as active participants (Calsamiglia 2003, Gülich 2003, Laslo et al. 2011). The described interaction always occurs with the traditional reader-in-the-text. The fictionalized reader supplies even greater evidence of the active role of the modern popular science reader. As Kranich et al. (2012) and Kranich (2016) point out, one of the major goals of a popular science author is to create a relationship with the reader that is "perceived as appropriate by the reader (based on his/her previous communicative experience and genre expectations)" (Kranich et al. 2012: 318). The expectations of the modern popular science reader would include having his/her influence acknowledged.

The authors who employ the techniques of the fictionalized reader quite often involve the reader in thought experiments through the use of presented discourse. Reader involvement in a text is not unusual on its own. For example, Kranich et al (2012: 318–319) note that "authors of English popular scientific articles tend to … simulate interaction with the reader. The reader is often addressed directly and drawn into the scenes described in the text." Using presented discourse of the reader as part of the interaction, however, is not a phenomenon that has received a lot of attention. Thought experiments that involve the reader are textual segments that attempt to translate abstract ideas into concrete examples through the application of the scientific principle being discussed to everyday situations. Thought experiments are often introduced in the form of games, narratives, or hypothetical scenarios that feature the reader as the main actant. In such instances, the reader is not a lone participant but interacts with other characters in the text. For example, Direct Speech of the reader can be used to form dialogues either with the author or with other characters in the books, who are, like the reader, given voice in thought experiments. The inclusion of the reader demonstrates that he/she is able to contribute to the discussion and application of scientific concepts. Consider an example from Greene (2011: 29):

6.15 Imagine it's a hot summer night and there's an annoying fly buzzing around your bedroom. You've tried the swatter, you've tried the nasty spray. Nothing worked. In desperation, you try reason. "This is a big bedroom," you tell the fly. "There are so many other places you could be. There's no reason to keep buzzing around my ear." "Really?" the fly slyly counters. "How many places are there?" In a classical universe, the answer is "Infinitely many." As you tell the fly, he … could move 3 meters to the left, or 2.5 meters to the right, or 2.236 meters up … Since the fly's position can vary continuously, there are infinitely many places it can be. In fact, as you explain all this to the fly, you realize that not only does position present the fly with infinite variety, but so does velocity.

By making the reader effectively a character in the story, the author eliminated Thompson's (2012: 80) concern for unpredictability of the real reader's reaction. The author is not dealing with a real-world reader any longer but has constructed a thoroughly fictional character. This fictional character takes on the task of explanation. Note that in the example from Greene (2011: 29) above, it is the fictionalized reader alongside the author who educates the fly: "As you tell the fly," "as you explain all this to the fly." In Greene's thought experiment, the fictionalized reader is also capable of drawing scientific conclusions and realizations, "you realize that ..." Such presentation of the reader contrasts sharply with the ignorant blank slate picture of the reader the dominant view of popularization projects.

Unlike presented discourse of scientists, presented discourse of the fictionalized reader has no real-world antecedent and in that is closer to presented discourse found in fiction, where, as Short (2012: 20) notes, the "idea of anterior vs. posterior discourse situations does not sensibly apply at all." The fictionalized reader also eliminates the need to have the author as a representative of the reader. Now there is a character with his/her own presented voice who fulfills this function. The author, thus, is free to speak *to* the reader without speaking *for* him/her also. Compare an example of the traditional reader-in-the-text as found in Thompson (2012: 80)—example 6.16.a—with the example of the fictionalized reader from du Sautoy (2011: 139)—example 6.16.b:

6.16.a Avoid fast food ... Fast food may be handy, but often is high in fat and calories.

6.16.b ... at some point, one of you will remove bars of chocolate so that the piles have 000,000, and 000 in them. Who does that? Your opponent always leaves an odd number of 1s in at least one of the piles, so it must be you that makes this move.

In both examples, the propositions attributed to the reader-in-the-text and to the fictionalized reader (respectively) are highlighted in blue. The reader-in-the-text is presented as part of narration, while the fictionalized reader is introduced through Free Direct Speech (Direct Speech version of the same question would be '"Who does this?" you might ask'.). The question posed in du Sautoy's (2011: 139) example is expressed by the fictionalized reader reacting to the events in the thought experiment. It is an acknowledgement of the reader's involvement in the thought experiment. While it could be argued that in Thompson's example (6.16.a) the writer is responding to a possible reader argument that "fast food may be handy," the acknowledgement of the reader is textually covert—that is the proposition attributed to the reader-in-the-text is concealed inside narration. du Sautoy (2011: 139), on the other hand, separates his own voice from that of the reader by using presented discourse.

Observed most often in thought experiments, the fictionalized reader serves a specific purpose. It aids in the explanation of scientific concepts. Thought experiments introduce material that could be challenging if presented in abstract terms. In such textual segments, the authors deliberately eliminate the abstract in favor of the specific. It is, therefore, understandable that they choose a more concrete fictionalized reader over the somewhat indeterminate reader-in-the-text. As those who explore evaluation in texts (see, for example, Hunston 2000, Martin and White 2005) point out, some of the interactional mechanisms that produce the reader-in-the-text can present a proposition not as a definite option but as one of the possible choices. Such open-endedness becomes undesirable in thought experiments where the focus is on the concrete.

Thought experiments deal with complex scientific concepts, and the authors try not to strain their resources by anticipating possible reactions or incorporating possible beliefs of the reader. They cannot afford a possible reader response that does not fit the fabric of the text. The fictionalized reader in addition to simplifying the writer's task by supplying exact reactions and eliminating guess work also helps the real reader. As Thompson (2001, 2012) mentions, if a real-world reader is not aligned with the attitudes and positions of the reader-in-the-text, the communication breaks down. However, the technique of the fictionalized reader allows the real reader to step back and experience explanations and arguments through another character in a text without breaking the communication line with the author in case of non-alignment.

The varieties of the presented discourse types that are assigned to the fictionalized reader suggest that this is a fully functional and well-developed phenomenon. As examples in Table 6.1 demonstrate, the types of presented discourse assigned to the fictionalized reader are comparable with the types of presented discourse of characters or actants found in large corpus studies in both fiction and non-fiction (see, for example, Semino and Short 2004). Both speech and thought presentation are observed. Another feature of the fictionalized reader that points to the significance and prominence of this technique is that it can be seamlessly integrated with presented discourse of other participants in the texts, forming dialogues. See example 6.15 above or Direct Speech example in the table below.

As the term "the fictionalized reader" suggests, the technique of fictionalizing the reader contributes to the similarities between popular science and fiction where the use of the second person pronoun creates the effect of involvement and suggests communication with real readers (Kacandes 2001, Mildorf 2016).

However, unlike in literary fictions, where the second person pronoun is used to bring the reader and the characters closer (and in some instances to merge them [see, for example, Fludernik 2011, Mildorf 2016]), the fictionalized readers in popular science can have their own narratives where they are the primary characters, who interact with other characters via a variety of presented discourse forms. Such passages could be regarded as similar to fanfiction where readers get a chance to apply their imaginations within the story world cre-

TABLE 6.1 *Discourse presentation types used to create the fictionalized reader*

Type of presented discourse	Example
Direct Speech	Imagine you work for the notorious film producer Harvey W. Einstein, who has asked you to put a casting call for the lead in his new indie, *Pulp Friction*. "How tall do you want him?" you ask. "I dunno. Taller than a meter, less than two ..." (Greene 2011: 152)
Narrator's Presentation of Speech Acts	... you ... tell the robotic cook in your kitchen to make breakfast and brew some coffee, and order your magnetic car to leave the garage and be ready to pick you up. (Kaku 2011: 354)
Free Direct Thought	Now you face a decision. How many actors should you have at the audition? You reason: If W. measures heights to a centimeter's accuracy, there are a hundred different possibilities between one and two meters. (Greene 2011: 152)
Direct Thought	Your first thought is, "Well, protons smash together, the Higgs comes out." (Carroll 2012: 166)
Narrator's Presentation of Thought Acts	Curled up under a warm duvet, just regaining consciousness but not yet having opened your eyes, you'll remember the Zaxtarian deal. At first it will seem like an unusually vivid nightmare, but as your head starts to pound you'll recognize that it is real ... (Greene 2011: 230)

ated by the author. They allow the reader to experience two common, according to Barnes (2015: 73), fanfiction scenarios: "inserting an idealized version of him/herself into the story world, where he or she can interact with favorite fiction characters" and engaging in "'what if' scenarios within the story world."

Thus, the fictionalized reader introduces a degree of fictionality into popular science.

The notion of a "degree of fictionality" appears in Skov Nielsen et al. (2015 a) and corresponds to the number of fictionalization techniques employed by the author, including the subject matter itself. The gradational view of fictionality is made possible by Skov Nielsen's et al. (2015a, 2015b) approach to fictionality as a communicative strategy. Of fictionalization mechanisms in non-fiction Skov Nielsen et al. (2015a: 65) say the following, "In uses of fictionality outside of generic fictions, a sender [author] does not transform nonfictional subject matter into something fictional but rather adopts a distinct communicative stance." In professional and popular science, the concept of the thought experiment often represents a communicative stance connected with fictionality. In fact, those who explore fictionality in professional science (see, for example, Barwich 2013, Toon 2012) focus on thought experiments and modeling as the most easily observable examples that supply evidence of the importance of imagination and fictionality to science.

Fictionality in popular science, however, is not employed for aesthetic purposes. It is inevitably linked to presentation of knowledge. By fictionalizing the readers, the authors help them process abstract scientific issues and make the texts not solely about science and scientists but also about the people who are reading the books. The presence of the fictionalized reader in the hypothetical passages of the texts ties the fictionality that this strategy produces to the fictionality found in professional contexts. In popular science books, the authors through the technique of the fictionalized reader give the actual readers a sample of the thought experiment process and so tie the act of reading popular science to the experience of doing professional science. Thus the fictionalized reader helps forge a unique connection not only between the reader and the author, but also between the reader and the scientific community.

Skov Nielsen's et al. (2015a, 2015b) approach to fictionality in non-fiction is strongly connected with the idea of hypothetical discourse, which accounts for the fictionalized reader. However, they do not explore this phenomenon but emphasize the general communicative role of fictionality, which the fictionalized reader in popular science clearly fulfills. Presented discourse that creates the fictionalized reader contributes to two important aspects of popular science. Firstly, it provides additional resources for the explanation of challenging scientific concepts, and secondly, supplies strong evidence in favor of presented discourse in popular science contributing to fictionalization of the texts. As far as the fictionalized reader's communicative properties are concerned, I suggest that this technique places presented discourse alongside such reader-construing elements as interactant pronouns, modalization, evaluation, etc.

The fictionalized reader reveals the power of fictionality as an interactional resource in popular science books and allows the reader more intimate access not only to the author but also to the scientific community. This technique also suggests that the use of fictionality extends beyond the creation of characters to the shaping of the reader. Numerous studies have been devoted to the examination of reader engagement strategies (for a sample of different approaches see Hyland 2001, 2005, 2009, 2010; Martin and White 2005; Lewin and Perpignan 2012; Fu and Hyland 2014), and all of them assume the reader to be an entirely real, albeit generalized, participant in a communication created by the text's author. According to such analyses, the author and the reader exist in opposition to the text's actants, who are entirely or to a certain degree products of the author's imagination or perception. Thompson's (2001, 2012) idea of the reader-in-the-text acknowledges the tentativeness of the reader, but it ends there.

My findings make it possible to take the idea further and show that in popular science books the authors fictionalize their readers to create concrete entities capable of reacting and communicating in a way that enhances the explanatory powers of the text. Presented discourse is used as the chief mechanism for the fictionalization of the reader. By fictionalizing the reader, the authors take more control of their audience than the current research suggests. Attributing some of the presented discourse to a reader (a specific, imagined reader-character rather than a generalized audience), produces for the real reader an effect of interacting with another character in the book. Thus popular science creates an interactive space where the readers can see themselves as part of the scientific community.

The fictionalized reader as an approach to reader engagement is significant since it demonstrates the willingness of the popular science authors to abandon the view of popular science that undermines the role and the knowledge of the public in favor of the supremacy of the scientific community. Bucchi (1998) and Myers (2003) criticize such a model of popularization, labeling it the "canonical view" and "dominant view" respectively. In general, as Kranich et al. (2012) and Mair (2006) show, modern popular science reflects cultural trends towards "the democratization of knowledge" (Kranich et al. 2012: 332). Some linguistic aspects of this change include "increased informality and colloquiality, as well as interaction" (Kranich et al. 2012: 331–332). The strategy of the fictionalized reader supports the heightened focus on interaction and suggests that modern authors of popular science books are on the path to a more inclusive and open model of knowledge dissemination.

Conclusion

A brief investigation of presented discourse outside the narratives of discovery provides a new outlook on the data collected from the narratives and helps confirm some of the propositions regarding presented discourse made in the context of the narratives of discovery. The analysis and comparisons undertaken in this chapter show that while such features as presentation of writing and the use of presented discourse for distancing are not prominent in the narratives, they are, nevertheless, present in popular science books. Presented discourse outside the narratives of discovery demonstrates that distancing occurs in connection with the information that is deemed incorrect or is not supported by the author. As such, distancing from certain voices in the text does not directly interfere with the general celebratory nature of presented discourse. In fact, by creating a contrast between acceptable and unacceptable knowledge, distancing contributes to the favorable presentation of those voices that project the right information. In light of the new evidence on distancing supplied by presented discourse outside the narratives, I suggest that the negative hypotheses introduced in the narratives can be seen as examples of distancing.

Another important feature of presented discourse in popular science that the narratives obscure but presented discourse outside the narratives brings to the forefront is presentation of writing. Outside the narratives of discovery, writing presentation is clearly identifiable as such and plays an important role, as it allows, for example, incorporating literary references.

The use of literary references and humor in presented discourse produces a more balanced portrayal of the scientists. Outside the narratives they are given opportunities to talk about their research in broader terms and to showcase the knowledge of, and the interest in, subjects that lie beyond their professions.

Perhaps the most significant finding of the chapter is the confirmation of the non-fiction and fiction-like qualities of popular science that presented discourse outside the narratives supplies. The analysis of presented discourse outside the narratives reveals that when it comes to forms and functions prototypically associated with either fiction or non-fiction, the alignment of form/function is distorted in the narratives but kept intact outside the stories of discovery.

Connected with the fiction-like features of popular science is a new technique for fictionalization that emerged as a result of the analysis undertaken in this chapter. It appears that the authors create fully fictional characters that represent the reader and express these characters through a variety of presented discourse types. I label this strategy "the fictionalized reader" and suggest that it is in the tradition of Thompson's (1995, 2001, 2012) idea of the reader-in-the-text. At the same time, I argue that the fictionalized reader is a more concrete entity that does not attempt to estimate the reader's attitudes but rather assigns them more aggressively. I see the fictionalized reader as a communicative strategy that supports Skov Nielsen's (2015a, b) view of fictionality as a tool of communication.

As such, the technique of fictionalizing the reader is designed to lift some pressure from the author and the real-world reader during the discussion of complex scientific concepts. For the author, the fictionalized reader creates an opportunity to project the reader's attitudes and reactions overtly. Unlike the reader-in-the-text techniques, which represent the reader covertly through the narrator's discourse, the use of the fictionalized reader produces its own discourse—presented discourse of the reader. The presence of a concrete reader-character in a text gives the author a chance to focus on the explanation without trying to embed the possible attitudes and reactions of the reader into the explanatory segments. For the real-world reader, the fictionalized reader offers a character with whom one can easily disagree without endangering the loss of communication with the author.

Overall, the analysis of presented discourse outside the narratives of discovery confirms the significance of presented discourse in the exposition of science for the public.

CHAPTER 7

Conclusion

The Role of Presented Discourse in the Formation of Positive Bias

As I have been demonstrating in the previous chapters, presented discourse in popular science is employed strategically to fictionalize and dramatize the events described. In fact, one would expect that including presented voices of scientists would have a somewhat "softening" effect—allowing the personalities and individual reactions to permeate the scientific facts. Quite often, when popular science is discussed as a genre (see, for example, Parkinson and Adendorff 2004), it is regarded as people-centered, rather than as texts that focus on theory or research methods (I have made similar claims in regards to popular texts' lack of focus on research procedures, see Pilkington 2017a). Parkinson and Adendorff (2004: 388) point out that popular science focuses "on people and what they say and think." This certainly suggests an important role for presented discourse. In fact, Parkinson and Adendorff (2004: 388) suggest that presented voices in popular science serve to create objectivity, as they introduce various expert points of view: "Authors of popular texts achieve the impression of objectivity by attributing ideas and utterances to the human participants." Moirand (2003: 181, 197), on the other hand, shows that multiple presented voices contribute to emotionality and not necessarily to objectivity though they do highlight various aspects of the issues discussed.

As my research shows, objectivity and multiple perspectives might not be the only goals presented discourse achieves in popular science. When Parkinson and Adendorff (2004: 389) claim that popular science texts present scientists as "personalities," they imply the focus on individual characteristics, idiosyncrasies, and emotions that are inevitably associated with the use of presented discourse. At the same time, a close look at presented voices of scientists demonstrates much more nuanced usage.

One assumption that many of those who mention presented discourse (but do not analyze it in detail) in their research make is that including someone's voice equals using a direct quotation (Direct Speech). Presentation of multiple voices, however, manifests in a variety of ways, many of which are indirect and thus less obvious to an observer who is not a specialist. Indirect forms of presented discourse (Indirect Speech and Thought, Narrator's Presentation of Speech and Thought Acts, for example) are widely used in non-fiction texts, including popular science. In popular science, however, as I demonstrate, they

© KONINKLIJKE BRILL NV, LEIDEN, 2018 | DOI: 10.1163/9789004365971_008

perform key roles usually assigned to direct forms (Direct Speech and Thought). The prominence of indirect forms and the reversal of functions of presented discourse forms in popular science make generalizations about the nature of presented voices problematic.

In fact, personal revelations through presented discourse are almost nonexistent in popular science. Presentation of thought of the members of the scientific community is devoted almost entirely to the presentation of scientific ideas; whereas, in other text types it is the go-to mechanism for revealing the inner worlds of the texts' actants. In popular science texts, any kind of emotionality and dramatization observed in presented discourse is directed towards scientific endeavors, not explorations of individual personalities. This is especially true for popular science narratives of discovery.

As such, the focus on the science rather than on the personalities of the scientists suggests that presented discourse brings a different contribution. In discussing scientific advancements, popular science aims to project an unapologetically positive image of the scientific community, and presented discourse is an ideal place to insert positive bias.

As I have demonstrated in the earlier chapters, the authors use presented discourse to show the members of the scientific community as relatable, creative, humorous, yet always engaged in doing science. If the scientists are assessed only on what their presented discourse reveals about them, they would appear unrealistically devoted to the pursuit of science at the expense of everything else. Such unwavering devotion, popular science texts show, is rewarded with exclusively positive research outcomes that rarely encounter any kind of complication that could not be overcome by hard work and dedication.

Dahl's (2015: 41) research into how popular science authors introduce certain perspectives (or frame their texts in a particular way) corroborates these observations. "From a writing perspective, framing may … be considered as a process that implies a strategic (conscious or subconscious) choice of angle (frame) by the text producer. The chosen framing implies selecting specific aspects of the issue/event at hand, making these particular aspects salient to readers." Choosing to include certain utterances or thoughts rather than others is a clear demonstration of an author's framing activity. However, the positive bias that is injected into popular science texts through presented discourse does not manifest only in the content of what the scientists say. It is also evident in the structures used around the presented utterances or thoughts—reporting clauses, the mechanisms used by the authors to integrate outside voices into their texts. Here is an example:

7.1 [Rous] barely admitted what he believed at some points: "It is perhaps not too much to say that [the discovery] points to the existence of a new group of entities which cause in chickens neoplasms [tumors] of diverse character."

KEAN 2012: 140

To make sure that the reader interprets Direct Speech of the scientist correctly, that is pays attention to the hesitation expressed via the hedge "perhaps not too much to say," the author projects the emotion of uncertainty via his own introduction. In this case, the author's evaluation mirrors that of the original speaker. However, in some cases it is not unusual to supply a positive evaluation of a neutral presented voice.

The positive bias introduced through a reporting clause belongs solely to the author. For example, Calsamiglia and Ferrero (2003: 149) write, "Citation means managing the words of others to convey and serve the purpose of the writer, giving a slant to what it said." However, to a reader of popular science the emotion and/or the evaluation introduced in presenting clauses appears to project the opinion of the scientist whose voice or thoughts is being reported. I discussed reporting clauses (Narrator's Representation of Speech/Thought—NRS/NRT) in detail in chapter 4 and demonstrated the emotionality that is injected into popular science texts through these devices. Here, I would like to reiterate the importance of these constructs in creating a positive presentation of science. As Calsamiglia and Ferrero (2003: 159) point out, "The image of the agent [a person whose discourse is being reported] is constructed and related through a specific reporting verb."

In popular science, especially in the text segments describing scientific discoveries, that image is nearly always positive, showing scientists as excited and passionate about their ideas. Dahl (2015: 55) observes that in certain cases reporting verbs "may … be interpreted as signaling association with the source's statement." That is, the assessment of the statement given by the author in the reporting clause mirrors the attitude of the speaker whose discourse he/she is presenting. The reader most often does not question this type of correlation, thus allowing the author to guide how a particular stretch of presented discourse is processed.

Such use of reporting clauses for introduction of bias is universal for all types of popular science texts; however, journalistic writing is less likely to include positive bias exclusively, while popular science in book form tends to focus on exclusively positive presentation. Analyses of popular science articles from newspapers, magazines, and journals (see, for example, Calsamiglia and Ferrero 2003, Dahl 2015, de Olivera and Pagano 2006) report instances of positive, negative, and neutral framing of presented voices. However, in my data collected from popular science books, the emotions associated with reporting clauses are always positive (see chapter 4), which allows me to classify reporting clauses as locations for positive bias. In fact, some reporting mechanisms add positive assessments to originally neutral statements, suggesting to the reader how to process the information in a positive way even if it was not delivered in that manner by the scientist whose ideas are being introduced.

The focus on the positive exposition of science and its practitioner also means that those voices that offer evaluation in addition to (or instead of) scientific explanations and facts get included more readily. The authors tend to frame outside voices in ways that would suggest evaluation rather than offer neutral presentation. It is an effective boosting mechanism. In the popular science texts I examined for this study, all the experiments, discoveries, and scientific advancements described were seen as providing contributions valuable to society. Not once was a discovery labeled as problematic or having undesirable social consequences. Even the issue of atomic science and the dropping of the atomic bombs on Japan were discussed in terms of advancements in cancer research. Such an approach is an example of social evaluation.

The idea of social evaluation of science was born in the eighteenth century when, according to Coppola (2016: 4), "scientific knowledge circulated far beyond the precincts of the experimentalists' laboratories," and as a result scientific facts "were stitched into and upheld the social fabric." For the first time, "It was not enough to merely ask what was true about nature; increasingly, one had to ask what were the implications of those truths—and of the particular practices of investigation and verification that produced those truths—to the wider culture in which they were located" (Coppola 2016: 4). In our time, popular science, at least in book form, has fully responded to the readers' demands for social evaluations of the information it presents and turned it into one more mechanism for the advancement of positive bias.

In addition to presentation of discourse, positive bias can be found in certain structural elements of popular science texts. For instance, popular science shows a great fondness for narrative, and narrative has built-in mechanisms for positive presentation of events. (See Appendix) On the other side, narrative structure also predisposes the authors to provide social evaluations of scientific breakthroughs.

As such, a lot of scientific discoveries are introduced as stories. The structural makeup of these stories embodies the double discourse of explanation and evaluation Moirand (2003) observed. The evaluative portion of the narratives of discovery (along with other structural elements) is part of a model identified by Labov (1972) in narratives of personal experience. (See Apeendix for more on Labov's model.) The fact that the evaluation element of popular science narratives has its roots in individual, personal experiences is not accidental. Myers (2003) demonstrates how social evaluation of scientific facts is connected with personal perceptions of science and its influence on individual lives.

The evaluation present in the narratives of discovery I analyzed can be divided into two types: social and scientific. Social evaluations reflect the ideas of Moirand (2003) and Myers (2003) and demonstrate how certain discoveries will affect the present society. Consider example 7.2.

7.2 This was intriguing, because scientists also believe that resveratrol may be responsible for the benefits of red wine and may explain the "French paradox." French cooking is famous for its rich sauces, which are high in fats and oils, yet the French seem to have a normal life span. Perhaps this mystery can be explained because the French consume so much red wine, which contains resveratrol.

KAKU 2011: 148

Kaku is writing about the discovery of the now-famous anti-aging chemical resveratrol, proposing that the French have been benefiting from it for a while, very likely without knowing what was going on. Now science has revealed the mystery. There is also an underlying current of advice or suggestion here: if this works for the French, it will likely work for you as well. In fact, the idea that quality red wine consumed responsibly can contribute to one's health is quite well-known by the general public. In Kaku's (2011: 148) example, we can see one of the points of its origin.

The social evaluation given here is unmistakably positive. There is a hint of the promise of a longer lifespan. A positive social evaluation is not a rarity in popular science, and especially in stories covering medical or biochemical discoveries. In fact, Molek-Kozakowska's (2016) investigation of popular scientific reports on the research on ageing suggests focus "on impacts/effects" and shows "that most frames for disease, ageing and death in popular science coverage work as vehicles for the celebration of medical science." My own data shows that evaluations included in the narratives of discovery are, for the most part, positive, no matter if a discovery is evaluated for social consequences or its contributions to scientific research.

The discovery of the graviton—"the quantum particle of the gravitational field," as described by Greene (2011: 81) supplies an example of a scientific evaluation. Consider example 7.3.

7.3 This is the most important accomplishment on string theory's resume and the reason it quickly soared to worldwide scientific prominence.

This is an evaluation not of potential benefits to society but rather benefits to the advancement of science. The reason the material on physics is given scientific and not social evaluation might have to do with the fact that the readers' demands for the social consequences of physics research are not as high at the moment as they used to be. As Molek-Kozakowska (2016: 50) notes, "Recent research on science-related coverage indicates a gradual reorientation away from physics, astrophysics, chemistry and engineering (popular at the time of nuclear proliferation, arms race and space exploration in the 1960s and 1970s) towards biology, medicine and biotechnology by the end of the twentieth century." The two types of evaluation given here represent research in biochemistry/medicine (example 7.2) and theoretical physics (example 7.3). According to Myers (2003: 269), "areas of specialist knowledge [of the public] are unevenly distributed; more people are interested in health and risk issues than in, say, algebraic theory or materials science." That is, fewer readers are likely to expect discussions of the social consequences of discoveries that are not likely to affect them personally. A recent development in string theory might seem less consequential than a medical discovery.

The data from the narratives I analyzed appears to support this idea, to an extent. Narratives covering medicine or biology and biochemistry are more likely to include social evaluations, while narratives dealing with discoveries in theoretical physics are more likely to be evaluated from the point of view of scientific benefits and further research. However, not all narratives covering theoretical disciplines follow this pattern, and scientific and social evaluations are distributed evenly among the narratives in my data with 50% of evaluations covering social consequences and 50% accounting for benefits to science.

It is common to think of mathematics as an entirely theoretical endeavor. It is not unheard of, for instance, to question the very physical existence of numbers (see, for example, Harris 2008). At the same time, du Sautoy not only reifies numbers in his popularizations but goes as far as to insist on the social repercussions of mathematical research. Thus, his story of how a three-dimensional tea bag came to be is an illustration of practical geometry and the power of mathematics to improve our lives. Consider example 7.4 that presents the social benefits of "the four-sided tetrahedron, or triangular-based pyramid":

7.4 Not only did it turn out to be efficient, but consumers thought the shape had a modern, funky feel to it. The new advertising certainly made a wel-

> come change to the troop of dressed-up monkeys that the company had relied on for years to sell its tea.
>
> DU SAUTOY 2011: 68

With this social evaluation, du Sautoy breaks the mold. His subject does not fall under the socially-crucial category, the way bioengineering or climate science would. Yet, by offering a social evaluation, he is demonstrating mathematics' potential for positive social effects. du Sautoy is obviously aware that the reader wants to know not only the facts of science but also see how the said science has already improved his life or might do so in the future.

The two sample evaluations introduced as examples 7.3 and 7.4 incorporate the voice of the author. My data suggests that this is by far the most common way to evaluate scientific discoveries. At the same time, presented discourse also plays a role: 30% of the evaluations include instances of either communicated discourse presentation or presentation of thought. The breakdown between social and scientific evaluations shows that presented voices and thoughts are equally distributed between them, with a slight preference for presentation of communicated discourse in the social evaluations. In general, as I have demonstrated in the previous chapters, presentation of thought in popular science tends to be used for introduction of scientific material.

The use of social evaluations in popular science narratives is a response to the readers' perception and processing of what science is and what it does. It might even appear that it is used strategically to promote positive bias since, by far, the majority of the evaluations are very positive. It is tempting to suggest that the authors are shaping their readers by controlling the form of the information they receive.

On the other hand, let us not forget that a modern reader is an active participant in the communication continuum among the scientific community, the popularizers, and the public. The relationship between the readers and the authors of popular science is quite complex and full of intricate details. There are several communication models that could be applied here to showcase the role of the reader. To my mind, the most effective one is Bell's (1984) audience design or a slightly modified version by Kacandes (2001). The framework is particularly apt for an analysis of popular science since it allows for an active recipient (reader). Kacandes' (2001: 154) uses the term "respondent" to designate an active participant in a communication. Bell (1984) and Kacandes (2001) show that just as a speaker/writer has the power to influence his listeners/readers, the listener/reader also exerts a certain amount of control. In other words, communication is a two-way street, be it between two people chatting face to face or between a news announcer and her audience.

For popular science this means that the shape of the information is to some extent controlled by the demands of the readers. Narratives are popular not necessarily because that is how the authors want to present their scientific facts but because that is what readers respond to the most. Positive presentation of science is also, studies suggest, a demand of certain audiences. The showcases of science's benevolence are indicators of a greater shift in style of non-fiction media in general. Molek-Kozakowska (2016: 51) draws attention to the relatively recent changes that occurred in journalistic writing: "Popular journalism involves not only shorter texts, more visuals and emotional headlines, but also more consumerism-oriented features, more personalization of reporting and domestication of abstract issues, more dialogical formats." All these features can also be observed in popular science, and, what is more surprising, in popular science books. Generally, popular science in book form is seen as more sophisticated and complex, catering to a readership that is more educated and dedicated (Hyland 2010: 118). On the other hand, as I have shown through analysis of presented discourse, there is a definite trend for emotionality. The phenomenon of the fictionalized reader creates a dialogic format—somewhat reminiscent of older popular science books which were constructed as dialogues. The presentation of discoveries as narratives of personal experience points to personalization, and social evaluations that come along with such narratives turn abstract research findings into potentially applicable solutions.

I should stress that all of these features are employed by popular science authors to promote and celebrate science. Why? One answer to this question I have already introduced above—science wants to promote itself. Another answer, however, has to do with readers' expectations. Celebratory discourse common in "science-exclusive outlet[s]" is a result of "readers' appreciation and engagement" with this type of information delivery (Molek-Kozakowska 2016: 52). While news programs and newspapers that cover science are more likely to report on scientific controversies and highlight negative developments, Molek-Kozakowska (2016: 52) suggests that this approach "would not be productive" for the public who reaches out to media devoted to popular science exclusively. In other words, readers who seek out popular science instead of encountering it among other information are expecting a positive, celebratory bias.

Modern popular science embraced the new terms of communication, where the readers are active participants shaping the content. In the process, the genre turned the readers' demands for social accountability and emotional accessibility to the advantage of the scientific community. For example, I see the phenomenon of social evaluation as part of a larger trend which connects the public to science on an emotional level. I discussed the importance of emotionality to popular science in chapter 2; here I wish to point out that the emotional connection is reader initiated.

As Myers (2003: 268) mentions, the lay public demonstrate again and again their, sometimes complete, lack of knowledge when it comes to the foundational facts of science. Yet, when seen as individuals rather than as a group, the lay people possess a variety of specific scientific knowledge pertinent to their individual areas of interest or problems. And as a group, the public knows not necessarily the facts of science but the aftereffects of scientific research. Science as a category is regarded in social and personal terms, not as a domain of objective truth.

The authors of popular science, therefore, have to balance their allegiance to the scientific community with the need to validate the readers' view of science. The result is texts that use emotionality and fictionality to introduce empirical results. The use of presented discourse in popular science reflects this trend very well.

In essence, the analysis of presented discourse introduced in this book is an acknowledgement of the social consequences of popularization. The importance of popular science as texts that raise awareness about the practitioners of science becomes evident in the connection between presented discourse and fictionality. This work strengthens the view of popularization that rejects the notion of a strict divide between professional and popular science. It attempts to showcase the fictionality of popular science as a unifying means between the two rather than as the point of divergence (as the traditional definition of fictionality suggests it might be).

In terms of linguistics, the analysis undertaken here confirms the importance of considering presentation of discourse outside the realm of literary fiction. As this work demonstrates, popular science as a borderline genre uses presented discourse in ways uncommon for fiction yet not conventional for non-fiction either. At the same time, the underlying idea behind the function of presented discourse in fiction remains intact in popular science. As Fludernik (1993: 64) summed it up, presented discourse aids in the reader's "creative reconstruction" of a world produced by the text. Fludernik's (1993: 64) statement is about "the fictional world" coming into existence in a real one. This book shows that through the presentation of discourse the opposite is also possible: a real world can take on fictional characteristics. Thus the real world of science becomes reconstructed as a forum where a lay reader can learn from, and about, a great number of great scientists.

Discourse presentation of scientists reconstructs the world of the scientific community as a human space but also projects back into it an image of a scientist the public wants to see. In that way, popular science can be regarded as a social construct that potentially shapes the personae of the very people it presents.

APPENDIX

Practical Applications of Research Findings

Popular Science as a Literary Genre

Popular science is many things. It is part of scientific discourse, as a forum for the presentation of science to the public; it is a non-fiction genre that combines information with entertainment (infotainment); it is an opportunity to voice and discuss cutting edge ideas that might not yet be ready to be accepted by the scientific community. And while these aspects of popular science are well-know and thoroughly addressed in multiple scholarly studies and writing manuals, another important side of popular science is usually left under-explored. Popular science, especially in book form, is a literary genre. As such it demands careful attention not only to the informational properties of the texts but also to the prose itself and to the characters.

Writing popular science has become increasingly important because of the pressure researchers and scientific institutions experience from policymakers, grant-providing agencies, and the public itself to make their findings comprehensible to lay audiences. The practical advice offered in this appendix is based on the research findings described in the main chapters of the book. This information is designed to serve as a reference for those authors who are interested in creating characters, using narrative techniques, and presenting the work of the scientific community in a positive light.

The Role of Presented Discourse in Science Communication to the Public

Studies in presented discourse undertaken in the twentieth century and the first decade of the twenty first, clearly demonstrate the importance of presented voices in shaping the public's understanding of science and the reaction to its advances. One of the key contributions of analyses of presented discourse is the identification of several problems with the way popular accounts of scientific activities have been written. For instance, the works of Calsamiglia, Ferrero, Hyland, Moirand, and Myers, to mention just a small fraction of the studies, supplied significant insight on the perception of the language of the scientific community by the public. For example, linguistic analysis of scientific (popular and professional) discourse denounced several potent myths about science. As Calsamiglia (2003: 141) puts it, linguistic studies of scientific discourse have the ability to "to remove many of the preconceptions—not to say prejudices—that people have about science." She continues, "Such ill-founded notions include the fact

© KONINKLIJKE BRILL NV, LEIDEN, 2018 | DOI: 10.1163/9789004365971_009

of considering science to be a neutral activity; the fact that science procures stable and eternally valid truths; the fact that the scientific community is licensed to account for natural, human and social reality; the fact of the 'hallowing' of scientific knowledge and its 'priests'; the fact of conceiving the linguistic representation of science as rhetoric free, maximally informative and transparent; the fact that science is a pure activity, out of touch with market forces and politics, etc." (Calsamiglia 2003:142).

The linguists working in the area of scientific discourse analysis have, in many ways, shone the light on the process of scientific knowledge construction and dissemination. They have also been relentless advocates for the lay public's access to the scientific community and have championed the public's role in negotiation of scientific knowledge. Myers and Moirand have been especially influential in highlighting the public's role not only in the consumption but also in the construction of popularizations. Myers (2003: 273–274), for instance, is responsible for identifying emotional connection and trust between the public and the scientists as vital qualities of popular scientific discourse; these elements have been independently pointed out as important to communication of science (see, for example, Fiske et al. 2002) and have become exemplary points in later research (see, for example, Sackler 2014). For more on the role of emotionality in popular science see chapter 2.

When one considers popular science from the point of presented discourse, several crucial features emerge, and all of them are connected to the idea of emotional engagement among the reader, the author, and the scientific community, in some way.

Direct and Indirect Speech, Narrator's Presentation of Speech Acts, Direct and Indirect Thought are among the most salient types of presented discourse of scientists when it comes to showing the members of the scientific community as relatable characters. Each of these presented discourse types is capable of producing striking rhetorical effects that tie popular science to literary fictions. It is crucial for any popular science author to know how to use them to his or her advantage.

Creating Engaging Characters

Direct Speech vs. Narrator's Presentation of Speech Acts: Construction of Dialogue

When it comes to demonstrating the human side of science, one of the best ways to showcase the members of the scientific community could be to present them in conversation. Traditional dialogue that is produced using Direct Speech, on the other hand, can potentially detract from the main message and can significantly slow down the pace of a text. Toolan (2001: 129) points out that any time an author uses Direct Speech of the characters, the action slows down. He calls it "a scenic slowing of pace." That happens because traditionally Direct Speech, even in non-fiction, is regarded as a dramatizing mechanism. Reporting verbatim what someone said would tell the reader more about that person than it sometimes would about the progression of the plot. Conversation and action are different things.

A conventional way to present dialogue is through Direct Speech. That is why popular science authors have to perform a delicate balancing act: on the one hand, they have to showcase their characters, the scientists, and on the other, they have to keep moving the narrative action along at a speed that the reader of a popularization expects. Another problem might be in the length common for narrative segments of popular science texts. The narratives found in popular science are not very long. They are not the end means but a mechanism to introduce background information that is necessary in order for the bigger issue to make sense. For example, Marcus du Sautoy, on the way to discussing the importance of prime numbers, takes a small detour to tell his readers the story of how zero was first discovered. In a narrative like this, an author might not always have enough room to include traditional dialogue, that is, employ Direct Speech. Below is an example of what is commonly recognized as a dialogue presented in writing:

A1 "Why don't you join us?" asked Charlea.
"I promised earlier not to interfere. It's time for you scientists to decide on your modus operandi. I'll just listen from here."
"No, join us."
"You mean you need another woman?"
"I do not mean that," replied C3, "but I'm starting to see how it might be useful to have a historical perspective."
DJERASSI 1994: 101

While useful, a dialogue is not the best method to relay information in a limited space. In fact, in his story about the Babylonians discovering zero, du Sautoy (2011: 23) does not use dialogue at all. And, of course, the kind of dialogue found in novels, does not exist in popular science writing. As I show in chapter 4, dialogue in popular science is realized through indirect forms of speech presentation, namely Narrator's Presentation of Speech Acts. That means that instead of giving the reader the dialogue in the original form (Direct Speech), the author simply has to say that a verbal exchange took place. The example above would then be converted to: Charlea asked me to join, explaining that they might need a historical perspective, but I declined. The conversion preserves the dialogic structure of question and response, but does not take nearly as much room and skips some of the minor details. The voices of the speakers, however, are preserved, yet they do not take over the action of the story.

In some cases, Narrator's Presentation of Speech Acts can be combined with other, less author-controlled, forms of presented discourse such as Indirect Speech and Free

Indirect Speech. Using these combinations is usually a good way to inject just enough of the personality of the speakers without slowing down the pace of the story. Direct Speech can also be used alongside Narrator's Presentation of Speech Acts. Particularly effective quotes could be made to stand out in a dialogue introduced using mostly indirect forms of presented discourse. Consider example A2, where Direct Speech is used to draw the reader's attention to the most important line in the dialogic exchange. Narrator's Presentations of Speech Acts (in blue) constitute the main body of the dialogue, while the instance of Direct Speech is used for emphasis.

A2 From the age of ten, Vera Rubin was fascinated by the stars. Her interest never waned, and when she applied to college it was only natural that she would seek to study astronomy ... At one point she spoke to a Swarthmore College admissions officer, who asked her whether she had any other interests. She admitted that she enjoyed painting. The admissions officer seized on that, asking, "Have you ever considered a career in which you paint pictures of astronomical objects?"

CARROLL 2012: 243

As Rubin's story continues, it becomes clear why Carroll (2012: 243) chose to showcase that particular line of dialogue: "... the question made an impression. She later recalled, "That became a tag line in my family: for many years, whenever anything went wrong for anyone, we said, 'Have you ever considered a career in which you paint pictures of astronomical objects?' " "

Using Narrator's Presentation of Speech Acts instead of Direct Speech when creating dialogue means emphasizing verbs. Note that the condensed dialogue in example A2 relies heavily on verbs to describe the kinds of speech acts that took place. First, Rubin "spoke" then the admissions officer "asked," which was followed with "admitted," and finally the verb "seized" prepared the reader for the main focus of the interaction—Direct Speech.

The four verbs encapsulate the core of the verbal exchange. The emphasis on verbs, while supplying a general idea of a dialogue, does not slow the pace of the overarching narrative of the book. Imagine what would happen if each of the exchanges expressed through one of the verbs was allowed to play out via Direct Speech. The interaction introduced in that manner would have taken more room than the author could afford—after all, Rubin's story (as are the majority of the stories in popular science) is not the main focus but an interjection that introduces background information.

Incorporating Narrator's Presentation of Speech Acts instead of Direct Speech gives the writer the freedom to conserve space and to name interactions rather than describe them. The interjection of Direct Speech, at the same time, gives a greater impression of the original voices and thus compensates for the lack of detail when the majority of the conversation is expressed via Narrator's Presentation of Speech Acts.

The suggestion to use indirect forms of speech presentation to express dialogue might seem counter-intuitive at first. After all, character conversations can introduce genuine emotion and personal perspective into a story. However, as I have demonstrated in the previous chapters, indirect discourse is capable of projecting emotionality and creativity just as well. It appears that the authors of popular science have found a way to use the resources available to non-fiction writers in such ways as to produce literary effects usually associated with fiction. Presentation of dialogue through Narrator's Presentation of Speech Acts is one of them.

Direct and Indirect Thought: Introducing Positive and Negative Hypotheses

In chapter 5, I demonstrated that the authors of popular science use presentation of thought to introduce discoveries and hypotheses into their narratives. Thought presentation is traditionally seen as the most intimate of all discourse presentation mechanisms. It is generally used to introduce the inner worlds of characters, to reveal insights so personal that even the characters themselves might not want to voice them. However, in popular science this inner-most way of discourse presentation ends up being used for what amounts to public announcements.

Similarly to the idea of expressing dialogue via indirect forms of presented discourse, the idea of using the most intimate to introduce the most important seems to originate with popular scientific prose. It is one more way to infuse a popularization with a personal connection between the reader and the members of the scientific community. Consider example A3.

A3 After his wife and daughter fell asleep for the day, Wilczek took a walk around the Fermilab grounds to think about physics ... On the same walk, Wilczek also came up with the idea for the "axion," a hypothetical low-mass cousin of the Higgs that is now a promising candidate for constituting the dark matter in the universe.

CARROLL 2012: 169

In the example above, the reader is given access to the scientist's thoughts. First, she is told that Wilczek was thinking "about physics," next she is a witness to the scientist coming up with a new idea, "Wilczek came up with the idea for the 'axion'." The scientific hypothesis of the axion being the dark matter particle is revealed to the reader as soon as the scientist had the idea. There is an instant connection here, an exclusive access, but not to the emotions of the character, rather to his scientific ideas.

In a written text just like in the real world, the ability to know what someone is thinking is limited. If in the real world only one person truly knows what he or she is thinking, on a printed page the narrator knows too. And it is at the discretion of the narrator that other characters and the reader get this information. The decision of the popular science authors to share presentations of hypotheses and discoveries in such a way that only the reader and not the other characters have access to them is a deliberate mechanism to establish intimacy and a personal connection.

Carroll (2012: 169) could have told the same story differently. At some point, Wilczek had to have made his hypothesis known. Carroll could have revealed the same idea as the scientist was relating it to his colleagues. However, the author chose to share the moment of the creation of the hypothesis. More so, quite often using thought presentation for introducing a hypothesis allows the reader to observe the thought processes of a scientist. Consider example A4. There are no revelations of Ørstead's inner world; instead the reader gets a chronologically accurate glimpse into the thought process that leads to the discovery:

A4 Ørsted had thought of a clever way to demonstrate the hypothetical connection between the two: He would build an electrical circuit, and then run the current next to a compass and see if its needle was deflected from true north by the running electricity.

CARROLL 2012: 121

Here we see how an idea for an experiment might be born. Ørsted's thought process reveals the steps necessary to prove a connection between electricity and magnetism. This short instance of thought presentation (Narrator's Presentation of Thought Acts followed by Free Indirect Thought) contains everything necessary to replicate this experiment.

Mastering the presentation of hypotheses and experiment designs as creative insights introduced through presentation of thought is an effective way to establish a connection between the reader and the scientists. It is also a mechanism that replicates for the reader some of the conditions for doing science. Fictionality (designing an experiment first in one's head or using a model before constructing a real experiment) is an inevitable part of scientific research in many disciplines. Including fictionality by using presentation of thought in a popular science text allows the reader to experience authentic scientific reality.

There is, of course, more to introducing hypotheses through presentation of thought than simply writing that someone had an idea or realized something. There are specific conventions on how to express different types of hypotheses. A hypothesis can be positive or negative. A positive hypothesis is the idea that finds empirical support. A negative hypothesis is a proposition that cannot be proven with experimentation. If an idea did not find experimental support or simply was dismissed after more careful deliberation, its description via presentation of thought would likely include the verbs "think" and "assume." If, on the other hand, the author is sharing a hypothesis that was later confirmed, the verbs "wonder" and "come up with an idea" will be used.

This observation is based on a relatively small sample of approximately 30,000 words and 193 occurrences of presented discourse, but it does indicate a writing strategy: positive and negative hypotheses should be introduced in different ways and the choice of reporting verbs should be consistent.

Indirect Speech: Showing the Creative Side of Scientists

Another insight for writers of popular science to come out of the research introduced in this book has to do with Indirect Speech. The overarching message of popular science is that the scientific community always comes up with brilliant ideas and that its members are wonderful and interesting people. That means the scientists are presented as infallible in their endeavors and at the same time as relatable human beings. One important aspect of showing the members of the scientific community in this way is demonstrating their creativity, and not only the creative approach to solving scientific problems but also the ability to use language in playful, unexpected, and elegant ways.

It is important for writers to use Indirect Speech rather than narration when using language creatively. Attributing a particularly unusual use of language, such as well-structured metaphor or a particularly effective analogy to a scientist and allowing him/her to take credit is essential. Such strategic uses of presented discourse show scientists as capable communicators who understand the needs of lay readers and are able to relate to them.

Quite often, the most striking figurative language could be found in Indirect Speech of scientists rather than in the segments narrated by the authors. This is not a result of the lack of talent on the part of the writers but is a deliberate strategy designed to endear the members of the scientific community to the reader. For example, Brian Greene (2012: 12) has Lemaître say that "galaxies were burning embers floating on a swelling sea of space." Ferris (1988: 256) and Bryson (2003: 139–140) attribute to Rutherford the analogy between alpha particles bouncing off a sheet of gold foil and a bullet hitting a piece of paper and falling down instead of easily going through. When Greene (2011: 39) writes about the papers by Gamow, Alpher, and Herman that provide theoretical evidence of the big bang, he introduces the expressive description of the conclusions as coming from the scientists by using Indirect Speech: "The authors concluded that we are all immersed in a bath of photons, a cosmic heirloom bequeathed to us by the universe's fiery birth." He could have easily presented the same figurative expression in his own voice: The authors' conclusions make me think that we are all immersed in a bath of photons, a cosmic heirloom bequeathed to us by the universe's fiery birth. Even though the metaphor very likely belongs to Greene, he is structuring his text in such a way as to attribute it to the scientists he is writing about.

There is some argument whether the information communicated through Indirect Speech originates with the speaker or with the one who reports it. Traditionally, indirect discourse, and therefore Indirect Speech, do not require absolute fidelity to the form of the original message but rather focus on the content. In cases when figurative language is used, the focus is inevitably placed on the form as well as on the content. Remember, however, that the bottom line of any successful popular science text is to relate scientific content in an understandable manner and to emphasize the positive qualities of the scientists, including their (and not necessarily the author's) linguistic creativity.

A popular science writer should not be afraid to attribute figurative language to scientists even if the original message may not have been very expressive. After all, one of the goals of a popularizer is to create relatable and interesting characters out of scientists. Fictionality in popular scientific writing manifesting through such character-creating mechanisms (among other means) has become the norm. Being creative does not always mean that the author takes full credit for a particular stretch of text. At times, it is more effective to attribute especially successful communicative strategies to the scientists that populate the stories.

Table A1 below provides a quick reference to the types and features of presented discourse that will help a popular science author create interesting and engaging characters out of the scientists he or she writes about.

TABLE A1 *Guide to presented discourse use in popular science*

Type of presented discourse	When to use	Benefits of use	How to use	Example
Direct Speech	Traditional dialogue	Introduction of details, emotions, background (used very sparingly in popular science)	Describe what a character said verbatim; use quotation marks	"I've got some free run-time on the Interactrex next week," Claire said, slapping a sheaf of papers on the desk between them ... "I've written a proposal." "Have you now?" Alan's drawl was too obviously designed to provoke to have any effect on Claire. "That's great, Claire." Ramon frowned at Alan before leafing through it (Rohn 2010b: 61)

Type of presented discourse	When to use	Benefits of use	How to use	Example
Narrator's Presentation of Speech Acts	Condensed dialogue	Brevity, emotion	Focus on verbs; tell the reader that a character spoke	In the pressroom afterward, reporters tried to get more comments from him, but he demurred ... (Carroll 2012: 185)
Direct Thought/ Indirect Thought	Introduction of scientific hypotheses	Intimate connection between scientists and readers; introduction of experiment design; explanation of thought processes; imitation of scientific reality	Use "wonder" and "came up with an idea" when introducing a positive hypothesis; use "think" and "assume" when reporting negative hypotheses	When Herschel observed planetary nebulae in which the central star was too dim to be seen, he assumed that they were globular star clusters. (Ferris 1988: 157)
Indirect Speech	Figurative use of language	Showing scientists as creative outside their primary domain	Attribute metaphors and analogies to scientists; include creative use of language as part of presented discourse	It was as if, he said, he had fired a fifteen-inch shell at a sheet of paper and it rebounded into his lap. (Bryson 2003: 139–140)

Using Narratives

In the world which demands and produces high volumes of information, the ability to organize knowledge effectively is essential. As I mentioned in chapter 3, narrative offers us, perhaps, the most efficient form of organizing and transferring information. We are hard-wired to learn from stories. It is no wonder that in recent decades such disciplines as communication of science to the public, philosophy of science, and sociology of science have been paying increased attention to narratives and their ability to transfer complex scientific concepts to lay audiences, to say nothing of the numerous books and workshops on using narrative designed to help scientists communicate their research.

Narrative structure can be easily reduced to a number of structural elements ordered in a particular way; however, there is much more to an expertly crafted story—especially if that story needs to be relatively short and serve as an explanatory detour.

Narratives are inevitably about people (or non-human characters who are endowed with anthropomorphic qualities), and it is vital to be able to handle the conversations and thoughts of the story's participants. That is where the ability to incorporate presented discourse becomes indispensable.

It is also extremely important to realize that in popular science, narratives are structural vehicles for ideology. They are used to promote the unapologetically positive image of science and its practitioners. I discuss this aspect of popular science throughout the book and in some detail in the Introduction, see chapter 1. Knowing how to handle narrative structure in order to produce a story that the scientific community would approve and would want to be told requires linguistic insight. An expert popularizer can easily manipulate the structural elements of a discovery story to respond to the readers' needs and to promote the positive image of science and the scientific community. This manipulation involves two major aspects: Explanation and Pattern Recycling along with Pattern Preview.

Explaining Science through Stories

Recent studies (see, for example, Reitsma 2010, Blanchard et al. 2015, Hermwille 2016) demonstrate that the explanatory and contextualizing abilities of popular science narratives appeal not only to the science-minded laymen but also to the decision-making social power structures such as grant-providing agencies or policy-creating institutions. This is because narratives have helped popular science to cross the boundaries of intellectual entertainment and become vital pieces of the technological and socio-economic spheres. A well-crafted narrative presentation of scientific research is likely to increase chances for funding and public appreciation. A lack of narrative representation results in ineffective communication of scientific research. As Olson points out in his book *Houston, We Have a Narrative: Why Science Needs Story* (2015: 12), "The costs of poor communication range from students getting bored with … science classes to the inability of the scientific community to deal with the growing antiscience movements for subject such as climate science, evolution and vaccination policy."

Narratives play such a crucial role in knowledge construction because they present information not as raw facts, but embed it in the context of human activity and emotion. The majority of popular science narratives include presented discourse. This feature alone indicates the focus on the people involved in addition to the scientific subject matter. The view of narrative as a context-rich environment that helps interpret scientific facts usually leads researchers to equate narrative with explanation (see, for example, Turney 2004b, Herman 2009, Reitsma 2010). Similar treatment of narrative as an explanatory mechanism can be found in less specialized sources targeting not linguists or narratologists but rather those who are looking for advice on effective communication of science (see, for example, Baron 2010, Olson 2015).

On the other hand, a narrative does not automatically equal explanation, and not all portions of a popular science narrative should be devoted to explaining. Most of the time, only certain parts of a popular science narrative should be used to explain, with other portions functioning as character vehicles or information delivery mechanisms. In order to learn how to include explanatory segments into a story correctly, it is necessary to understand the overall structural makeup of a traditional narrative.

Narrative Structure: Labov's Model in the New Century

Labov's (1967, 1972, 1997) work concerns basic narrative structure. He is famous for analyzing narratives of personal experience and pointing out the universal structure such texts (oral and written) follow. Since Labov's (1972) work was published, a number of researchers have confirmed his findings and extrapolated from them that Labov's framework is essentially applicable to all types of narratives, not only to those that tell personal stories.

In fact, Labov's (1972) model (with modifications) has served as a springboard for structural analyses of literary narratives, narratives of children, and several other types of personal narratives not examined by Labov (see, for example, Baerger and McAdam 1999, Berman 1997, Fleischman 1997, Peterson and McCabe 1983). So far, the value of the model has been in its universality; it is capable of exposing the underlying narrative structure of multiple text types. However, Labov's (1972) framework for narrative analysis can do more. When applied to certain segments of popular science texts, it not only supports the claims for narrativity but also helps identify a structural element that is peculiar to popular science narratives of discovery. Not surprisingly, this structural element is Explanation, and it accounts for the scaffolding approach to information that so many popular science texts take. I will return to it shortly, but before discussing additions and modifications to the classical narrative structure, it is necessary to examine briefly Labov's (1972) original model.

In 1967, Labov and Waletzky introduced a structural framework for analyzing oral narratives of personal experience. In 1972, through a case-study, Labov perfected this model. The original Labovian approach suggests that narratives of personal experience can be parsed into several sections that correspond to the following narrative elements:

- Abstract (an optional brief summary of a narrative to come)
- Orientation (introduction of the story participant and the setting of a narrative)
- Complication (a difficulty to be resolved; the main body of a narrative)
- Evaluation (the speaker's assessment of a narrative), Result (the resolution of the aforementioned difficulty)
- Coda (the element that transitions the speaker and the listener from the past events of a narrative to the present).

Labov (1972: 369) suggests that a complete narrative contains these elements and presents them in the order listed above. Ultimately, Labov's (1972) model shows how to shape information into a story. Here is an example of a popular science narrative of discovery analyzed through Labov's model:

A5 "Galileo and Pendulum"

Abstract
Why Are Pendulums Not As Predictable As They First Appear?

Orientation
It was Galileo, the master of using math to make predictions, who first unlocked the secret of what makes a pendulum tick. The story goes that when he was 17, he was attending Mass at the cathedral in Pisa.

Complication
In a moment of boredom, he stared up at the ceiling, and his eyes fell on a chandelier that was swinging gently in the breeze blowing through the building. Galileo decided to time how long it took the chandelier to swing from side to side. He didn't have a watch (they hadn't been invented yet), so he used his pulse to keep track of the swing.

Result
The great discovery he made was that the time the chandelier took to complete one swing did not seem to depend on the size of the swing.

Explanation
In other words, the time of the swing essentially doesn't change if you increase or decrease the angle of swing. (I put the word essentially in there to indicate that if we dig a little deeper, things get slightly more complicated.)

Result (cont.)
When the wind blew harder, the chandelier swung through a larger arc but took the same time to swing as when the wind dropped and the chandelier was hardly moving at all.

Evaluation
This was an important discovery and resulted in the swinging pendulum being used to record the passage of time.

> Coda
> But what does the time of the swing depend on, and can we predict whether and how the swing will change if the weight is increased or if the pendulum is made longer?
>
> DU SAUTOY 2011: 226

The Orientation, the Complication, and the Result in this narrative represent what Labov (1972) called a "temporal sequence"—a chronological order of events. The Abstract and the Coda could be part of the temporal sequence as well, but they are not in this narrative. The Evaluation is always positioned outside the chorological order, and sometimes it can contrast with the body of the narrative by using the present tense instead of the past. In addition to the elements of Abstract, Orientation, Complication, Evaluation, Result, and Coda popular science narratives of discovery contain (as example 6.5 does) an additional element, not present in Labov's model, "the Explanation." Notice that it is a specific segment of the narrative (not the whole story), and it has identifying grammatical features.

Similarly to the Evaluation, the Explanation may initiate a shift from the past tense of the body of the narrative to the present tense, as it does in the example above. The transition from the Result to the Explanation is easy to detect precisely because of this tense shift. The Explanation element, in addition to following another narrative element, can easily "interrupt" a section, creating a split in the Result or the Complication, for example.

The Explanation is part of what Labov (1972: 369) considers a "secondary structure" of a narrative—that is a portion of a narrative that relates to the listener/reader the information other than the temporal sequence of events. Together with the Evaluation, it supplies the information that is essential for the processing of the narrative by the reader. The Explanation and the Evaluation provide the context necessary for the reader of popular science. In some ways, these two elements embody the popularizing techniques.

The Explanation, as a structural element, should be positioned strategically at a point where the author anticipates any kind of confusion in the reader. It should be confined to a particular section of a narrative and should not dominate the story. For instance, in the narrative above (example A5), the Explanation occupies only 15% (41 words out of 267 words total) of the story. In fact, no single element ever occupies more than about 30% of the total length of the narrative.

Any narrative element that is not part of the temporal sequence, that is, not contributing to the action, suspends the story. So keeping the Explanation short is essential. An expert popular science author would want to make sure that his or her reader follows along and is not lost among unfamiliar terminology and concepts. One way to achieve this is to use explanatory waves in addition to the main Explanation. Explanatory waves usually manifest as short definitions or asides found within the elements representing the temporal sequence. Consider example A6 with a short definition (in blue) inserted into the Result section of the narrative:

A6 Suddenly, dots arranged in pentagons appeared, and that allowed them to home in on one of the more interesting of Plato's dice: the icosahedron—the shape made of 20 triangles with five triangles meeting at each point.

DU SAUTOY 2011: 69

In an especially complex narrative, an author might need to go beyond one dedicated explanatory segment and several explanatory waves. Sometimes it is necessary to employ multiple Explanations. That means positioning the Explanation element not only in its usual place—after the Result, but also following or interrupting other narrative sections. These separate explanatory sections are different from the explanatory waves because they interrupt the temporal sequence by creating a tense shift; explanatory waves are located within the temporal sequence and do not initiate tense shifts.

Examples A7.a, A7.b, and A7.c demonstrate how multiple Explanations interject into a temporal sequence and suspend the narrative action. All of these segments are taken from the same narrative. There are three explanatory elements dispersed throughout the story: one located in the usual position immediately after the Result, one appearing right after the Abstract, and one splitting the Complication. Each time the narrative shifts to the Explanation, the tense changes from the past simple to the present simple.

A7.a Result

The first rule was that soap films always meet in threes at an angle of 120 degrees. The edge formed by these three walls is called a Plateau border in his honor. The second rule was about the way these borders can meet.

Explanation

Plateau borders meet in fours at an angle of about 109.47 degrees ($\cos^{-1}$-1/3, to be precise). If you take a tetrahedron and draw lines from the four vertices to the center, you get the configuration of the four Plateau borders in foam. So the edges in the bulging square at the center of the cube wire frame actually meet at 109.47 degrees.

A7.b Abstract

Five years later, a pencil-and-paper proof of the double-bubble conjecture was announced. It actually proved a more general conjecture:

Explanation
if the bubbles do not enclose the same volume, but rather one is smaller than the other, then the bubbles fuse together so that the wall between the bubbles is no longer flat but bent into the small bubble. The wall is part of a third sphere and meets the two spherical bubbles in such a way that the three soap films have angles of 120 degrees between them.

A7.C Complication
While he was doing research into the effect of light on the eye, he stared at the sun for half a minute, and by the age of 40, he was blind. Then, with the help of relatives and colleagues, he switched to investigating the shape of bubbles. Plateau began by dipping wire frames into bubble mixture and examining the different shapes that appeared.

Explanation
For example, when you dip a wire frame in the shape of a cube into the mixture, you get 13 walls that meet at a square in the middle. This "square," however, isn't quite a square—the edges bulge out.

Complication
As Plateau explored the various shapes that appeared in different wire frames, he began to formulate a set of rules for how bubbles join together.

DU SAUTOY 2011: 70–74

The Explanation is similar to Labov's (1972) Evaluation in the sense that in many narratives it functions as a suspending device for the story and could be excluded without changing the temporal order of events. At the same time, the Explanation is closely interwoven into the fabric of a narrative, and its exclusion will impact comprehension. To popular science narratives of discovery, the Explanation might be as essential as the Evaluation is to narratives of personal experience. Labov suggests that without the Evaluation, such narratives lose their purpose. Moreover, Labov (1972: 372) demonstrates that under certain circumstances, the Evaluation in the narratives of personal experience is essential in order to elicit the right emotional response from the audience and to keep them interested. The Explanation functions in a similar way.

Without the Explanation, the narratives of discovery lose much of their meaning since a lay reader might not be equipped to understand the very subject matter of a discovery. In a way, the Explanation element of the narratives of discovery facilitates the communication between the author and the reader (Pilkington 2017b: 12–14).

The Explanation is a perfect place for the author to interact with the reader. Since it takes the reader away from the narrative action, it gives the author an opportunity to talk to the reader directly and even to engage her in scientific experiments. Other than the tense shift, the Explanation allows for a switch in the narrative focus from the scientist to the reader. Look to examples A7.a and A7.c above. By stepping out of the chronologically ordered events of the story, the author is able to address the reader directly and ask him/her to engage the imagination to help comprehend the events described. du Sautoy is using the pronoun "you" to draw the reader into the hypothetical activities used to create the Explanation. The Explanation becomes a thought experiment; however, this time, it is not the scientist who experiments but the reader. In some small way, this is an opportunity for the reader to experience what it is like to do science.

It is possible to take the interaction with the reader a step further and to merge the identities of the author and the reader, allowing the reader to take part in a personal experience of the author. Brian Greene does that in his book *The Hidden Reality*. In the Explanation section of a narrative about duality, Greene (2011: 109–111) uses the interpersonal pronoun "you," indicating the reader, while relating his own personal experience. By doing so he creates an illusion of the reader seeing the same things as the author did. He thus transforms his own observations into an experience of the reader, helping him/her connect with the material on a personal level. See example A8.

A8 Explanation

I recently encountered a splendid graphic that from close up looks like Albert Einstein, with a bit more distance becomes ambiguous, and from far away resolves into Marilyn Monroe … If you saw only the images that come into focus at the two extremes, you'd have every reason to think you were looking at two separate pictures. But if you steadily examine the image through the range of intermediate distances, you unexpectedly find that Einstein and Monroe are aspects of a single portrait. Similarly, an examination of two string theories, in the extreme case when each has a small coupling, reveals that they're as different as Albert and Marilyn. If you stopped there, as for years string theorists did, you'd conclude that you were studying two separate theories. But if you examine the theories as their couplings are varied over the range of intermediate values, **you** find that, like Albert turning into Marilyn, each gradually morphs into the other.

Overall, the presence of the Explanation demonstrates a structural complexity of popular science narratives. Not only does it supply an additional element for the narrative structure, but it also reveals the possible re-organization of the elements that Labov (1972; 1997) did not anticipate. In addition to the interruptions of the temporal sequence, the Explanation can also be inserted inside the Evaluation, creating a complex cluster of scaffolding elements. Such scaffolding clusters are most likely to appear at the end of a narrative, following the Result. The scaffolding cluster created by the

Explanation and the Evaluation does not mean that the Explanation is not present as a separate element elsewhere in the narrative. Example A9 taken from Michio Kaku's (2011: 148) narrative about the discovery of Resveratrol illustrates a scaffolding cluster:

A9 Evaluation

This was intriguing, because scientists also believe that resveratrol may be responsible for the benefits of red wine and may explain the "French paradox."

Explanation

French cooking is famous for its rich sauces, which are high in fats and oils, yet the French seem to have a normal life span. Perhaps this mystery can be explained because the French consume so much red wine, which contains resveratrol.

Evaluation

Scientists have found that sirtuin activators can protect mice from an impressive variety of diseases, including lung and colon cancer, melanoma, lymphoma, type 2 diabetes, cardiovascular disease, and Alzheimer's disease, according to Sinclair. If even a fraction of these diseases can be treated in humans via sirtuins, it would revolutionize all medicine

It might appear that the Explanation complicates the narrative by destroying the linear chronological structure. However, in the case of popular science narratives, the structural complexity actually promotes comprehension (Pilkington 2017b: 19). Making a narrative more complex structurally allows the authors extra opportunities for creating the context for the information presented. The value of a narrative is not only in the familiar, linear, chronological sequence of events but also in its ability to connect with the reader and to personalize the story being told. The Explanation allows popular science authors to do just that.

Labov's (1972) framework is invaluable for helping identify the element of Explanation. At the same time, the reach of the model is limited as it does not allow the exploration of narratives beyond their structural makeup, and there is more to a good popular science story than a chain of properly ordered narrative elements. Other than a solid structure, an effective popular science narrative possesses a certain degree of ideology. Its goal is to reassure the reader that the scientific community is nearly infallible and its activities are directed to the greater good of humanity.

Pattern Recycling and Pattern Preview: Hoey's Model as a Detector of Positive Bias

Professional scientific writing is often seen as objective and agenda-free, and by extension popular science has been regarded in a similar way. However, this is not so. As

Harré (1994) points out, professional scientific writing focuses exclusively on positive outcomes of laboratory procedures. As early as 1620, in his book *New Organon*, Francis Bacon acknowledged the influence of a well-told story on the scientific community and their activities. "According to Bacon, the laboratory is uncomfortably akin to the stage insofar as audiences for both scientific and dramatic performances are motivated by an appetite for 'elegant narratives,' and not the messy and inconclusive 'true stories from nature'" (Coppola 2016: 2). Popular science can also be accused of the pursuit of "elegant narratives" sometimes at the expense of the truth. Popularizations successfully accomplish the goal of presenting the scientific community in an almost exclusively positive light. The work of Hoey (2001) on narrative dynamic and textual patterns supplies linguistic insight into the celebratory discourse of popular science and shows how positive bias is produced.

Similarly to Labov, Hoey (2001) identifies several structural elements of a narrative. He calls them the Situation (which is essentially an introduction, an opening move of a narrative), the Result, and the Evaluation. However, Hoey (2001) takes a more complex approach to what Labov (1972) named the Complication and suggests not one element but several subtypes. Accordingly, a Complication could represent a problem, a goal, a gap in knowledge, an opportunity, or a desire. Each type of Complication has a corresponding Result type: thus a problem Complication would have a Result that represents a solution; a goal Complication has a Result that represents the achievement of that goal; for a gap-in-knowledge Complication, the Result offers the filling of the gap; a Complication that represents an opportunity has a Result that demonstrates that the opportunity was taken, and a Complication that arouses desire results in its fulfillment.

Hoey (2001) suggests that these combinations describe the most popular ways to organize stories in Western cultures and proposes that each of the Complication/Result combinations represents a narrative pattern. These patterns are:

- the Problem-Solution pattern
- the Gap in Knowledge-Filling pattern,
- the Goal-Achievement pattern
- the Opportunity-Taking pattern
- the Desire Arousal-Fulfillment pattern.

Each pattern is triggered by its own lexical signals. As Hoey (1983) points out, a lexical signal is a type of signaling vocabulary that helps the reader interpret the relationships between sentences or groups of sentences in a given text. According to Hoey (2001: 124) certain words trigger certain expectations of how a narrative will proceed or, in other words, what pattern it will follow. Hoey (1983: 63) writes, "Lexical signaling can take the form of a sentence, clause or phrase and incorporates either one or more typical signals or an evaluation." For example, Hoey (2001: 124–125) proposes that such "negative evaluation items" as "problem" or "unable" are signals of a narrative that presents a problem to be solved. When a reader sees these words, he/she may expect the narrative to describe what was done about this problem. A writer who uses such words is expected to offer not only the statement of a problem but also an account of the actions taken to resolve it. Hoey (2001) refers to the first lexical signal of a narrative as a trigger. The trigger determines the type of pattern a narrative will follow.

While each of Hoey's patterns is different and will produce different stories, there are overarching similarities. First of all, each pattern follows the negative/neutral to positive progression: from a problem to its solution, from a desire to its fulfillment, from not knowing to gaining knowledge, and so on. While a story may begin on a negative (problem, gap) or a neutral (opportunity, desire, goal) note, it inevitably ends on a positive one. There is not a single pattern that presupposes a negative outcome.

Hoey (2001: 130) argues that whenever a narrative has a negative outcome, "the pattern has to recycle." In other words, if a problem has not been solved or the desired knowledge not obtained, an author has to keep telling the story until a positive outcome can be reached.

A10 "Chaos Theory"

Setting of the Situation/Introduction

In 1885, King Oscar II of Sweden and Norway offered a prize of twenty-five hundred kroner for anyone who could establish mathematically once and for all whether the solar system would continue turning like clockwork, or whether it was possible that at some point, the earth might spiral away from the sun and off into space.

Problem

Poincaré thought he could find the answer, and he began to investigate. One of the classic moves that mathematicians make when they are analyzing complicated problems is to simplify the setup in the hope that it will make the problem easier to solve. Instead of starting with all the planets in the solar system, Poincaré began by considering a system with just two bodies. Isaac Newton had already proved that their orbits would be stable: the two bodies just travel in elliptical orbits around each other, forever repeating the same pattern. From this starting point, Poincaré began to explore what happens when another planet is added into the equation. The problem is that as soon as you have three bodies in a system, for example the earth, moon, and sun, the question of whether their orbits are stable gets very complicated-so much so that it had stumped even the great Newton. The problem is that now there are 18 different ingredients to combine in the recipe: the exact coordinates of each body in each of three dimensions, and their velocities in each dimension. Newton himself wrote that "to consider simultaneously all these causes of motion and to define these motions by exact laws admitting of easy calculation exceeds, if I am not mistaken, the force of any human mind."

What was Done
Poincaré was not daunted. He made significant headway, simplifying the problem by making successive approximations to the orbits. He believed that if he rounded up or down the very small differences he found in the positions of the planets, it wouldn't affect his final answer too much.

Solution
Although he couldn't solve the problem in its entirety, his ideas were sophisticated enough to win him King Oscar's prize.

Problem
However, when Poincaré's paper was being prepared for publication, one of the editors couldn't follow Poincaré's mathematics and raised a question. Could Poincare justify why making a small change in the positions of the planets would result in only a small change in their predicted orbits?

What was Done
While Poincaré was trying to justify his assumption, he suddenly realized he'd made a mistake. Contrary to what he'd originally thought, even a small change in the initial conditions—the starting positions and velocities of the three bodies—could end up producing vastly different orbits: his simplification didn't work. He contacted the editors and tried to stop the paper from being printed, because publishing an erroneous paper in honor of the king would have caused a furor. The paper had already been printed, but most of the copies were gathered together and destroyed.

Final Outcome/Discovery
It all seemed a huge embarrassment. But as often happens in mathematics, when something goes wrong, the reason it goes wrong leads to interesting discoveries. Poincaré wrote a second, extended paper explaining his belief that very small changes could cause an apparently stable system to suddenly fly apart. What he discovered through his mistake led to one of the most important mathematical concepts of the last century: chaos theory.

DU SAUTOY 2011: 229–230

This narrative is a classic example of pattern recycling. However, notice that the recycling trigger—the solution to the initial problem—is not fully negative, "he couldn't solve the problem in its entirety," and the evaluation that follows it is positive, "ideas were sophisticated enough to win him King Oscar's prize." The pattern recycling is prompted not only by the partially resolved initial problem but also by the interference of another actant, "one of the editors couldn't follow Poincare's mathematics and raised a question." In this instance, Narrator's Presentation of Speech Act functions as an indicator of a new problem that the rest of the narrative has to resolve. The choice of Narrator's Presentation of Speech Act (as opposed to Direct or Indirect Speech) is not accidental. It places the focus on the action—raising the question—and not on the exact wording, which is not as relevant. The fact of the interference is what triggers the discovery in this case.

Pattern recycling can also take a different form. In a limited space, it can be accomplished through partial recycling of a pattern, where an initial negative outcome is followed immediately by a positive resolution. Consider example A11:

A11 "Zero"

Setting of the Situation/Introduction
The Babylonians came close to discovering a very important number in mathematics: 0.

Problem
If you wanted to write the prime number 3,607 in cuneiform, you had a problem. This is one lot of 3,600—or 60 squared—and 7 units, but if I write that down, it could easily look like one lot of 60 and 7 units—still a prime, but not the prime I want.

What was Done
To get around this, the Babylonians introduced a little symbol to denote that there were no 60s being counted in the 60s column, So 3,607 would be written as: [picture]

Negative Outcome
But they didn't think of 0 as a number in its own right. For them, it was just a symbol used in the place-value system to denote the absence of certain powers of 60.

Positive Outcome

Mathematics would have to wait another 2,700 years until the seventh century AD, when the Indians introduced and investigated the properties of 0 as a number.

DU SAUTOY 2011: 23

In this example, pattern recycling is triggered by a problem at a conceptual stage of the discovery process, "But they didn't think of 0 as a number in its own right." This negative resolution should, according to Hoey (2001: 130), initiate essentially a new narrative—one with a positive outcome. However, this is not what the author chooses to do. Instead of recycling through the whole pattern and supplying another narrative, du Sautoy chooses to skip to the positive result.

In addition to pattern recycling, Hoey (1983) suggests another possible narrative development that in popular science is used to promote positive presentation of scientists and their activities. It is not uncommon for a narrative to present a summary of a pattern (for example, name the problem and immediately indicate that it was solved) and then continue to explain how the result was reached. Hoey (1983: 138) calls such an arrangement of narrative elements a "the Preview-Detail" pattern. Consider the following example:

A12 String Theory and Black Holes

Situation

Nevertheless, one major advance has illuminated a related aspect of black holes. As I will discuss in Chapter 9, the work of Jacob Bekenstein and Stephen Hawking in the 1970s established that black holes contain a very particular quantity of disorder, technically known as entropy. According to basic physics, much as the disorder within a sock drawer reflects the many possible haphazard rearrangements of its contents, the disorder of a black hole reflects the many possible haphazard rearrangements of the black hole's innards.

Problem

But try as they might, physicists were unable to understand black holes well enough to identify their innards, let alone analyze the possible ways they could be rearranged.

Solution—general (preview)

The string theorists Andrew Strominger and Cumrun Vafa broke through the impasse.

Solution—specific (detail)
Using a melange of string theory's fundamental ingredients (some of which we will encounter in Chapter 5), they created a mathematical model for a black hole's disorder, a model transparent enough to enable them to extract a numerical measure of the entropy.

Result and Evaluation
The result they found agreed spot-on with the Bekenstein-Hawking answer. While the work left open many deep issues (such as explicitly identifying a black hole's microscopic constituents), it provided the first firm quantum mechanical accounting of a black hole's disorder. The remarkable advances in dealing with both singularities and black hole entropy give the community of physicists well-grounded confidence that in time the remaining challenges of black holes and the big bang will be conquered.

GREENE 2011: 98

Right after the problem has been identified, the reader knows that there exists a solution. This is the preview portion of the pattern. The next narrative step shows how the solution was reached, and the final step evaluates and reaffirms the solution as a positive outcome.

A preview indicates an important angle. By placing the outcome at the beginning of the narrative, the author dismisses the possibility that a problem can be left unresolved. Using this type of narrative flow, the authors respond to their writing task in a way that results in a structure that is different from the actual temporal order of the events that took place. The goal of such re-ordering is to eliminate any doubt as to the positive outcome of a narrative—a mechanism employed to engineer positive bias into the narrative structure.

The preview-detail organization is popular but not universal in popular science narratives. It occurs in about 50% of the Problem-Solution and Gap in Knowledge-Filling narratives, which represent the two most common patterns for popular science.

Table A2 offers a comparison between Labov's (1972) and Hoey's (2001) models for easy reference.

TABLE A2 *Narrative structure according to Labov and Hoey*

Narrative Element	Labov's Model	Hoey's Model
	Abstract	Situation
	Orientation	
	Complication	Problem, Gap in Knowledge, Goal, Desire, Opportunity
	Evaluation	
	Explanation	
	Result	Solution, Filling, Achievement, Fulfillment, Taking
Main Function in Popular Science	Identification of explanatory segments	Identification of positive bias through pattern recycling and preview

Conclusion

Popular science as a genre of literature has developed its own conventions for the use of presented voices and narrative structures. Linguistic insights can guide popular science authors to handling these conventions successfully. However, the majority of the literature on the subject of popular science writing (various how-to manuals) lack the in-depth knowledge when it comes to explaining these concepts.

For example, the understanding of presentation of discourse is rather marginal in many of the how-to guides, with the notion of presented discourse equated with quoting. At the same time, the works of Meredith (2010: 18) and Sackler (2014: 8–10) acknowledge the vital role scientists play in making science relatable. They also demonstrate a concern that the public does not always see the members of the scientific community as "warm" (Sackler 2014: 10). The recognition and the concern, however, do not result in more reliable information on the presentation of the voices of scientists. A piece of advice along the lines of "Include the researcher's name, affiliation, and a sentence or so about the work and how it fits into the topic" before quoting is easy to follow (Meredith 2010: 211). But it does not in any way help present the scientist as a warm, interesting, or in any way retable person. This is a cardboard cut-out, not an engaging character. There is much more to presented voices of scientists than quoting.

The approach to the task of writing popular science that I present in this appendix is different from the existing methods because I argue that the features usually attributed to narration and therefore to the author (explanations, presentation of hypotheses and discoveries) should be fulfilled by presented voices of scientists. The reason for sharing the burden of delivering information between the writer and the presented voices is grounded in the issue of trust. Myers (2003: 273) wrote that popularization, among other matters, "involves the active construction of believable … identities" and that "people assess messages … in terms of … trust." Sackler (2014) laments that "Trust has been a largely neglected topic in the science of science communication" (8). In general, though, he echoes Myers' (2003) message in asserting that "people trust those who they think are like themselves." Both agree that trust implies not only professional competence but also the ability to relate a message effectively. Using multiple voices of the scientific community rather than the one voice of an author is more productive. It shows scientists not as abstract entities in the margins of the social construction of scientific knowledge but as active participants in this process who care about their work being understood and accepted by the public.

In general, the close connection between popular and professional scientific writing is not always beneficial to those who want to popularize. This goes beyond the idea that scientists who venture into the field of popularization must undergo drastic changes in their attitudes toward the lay public (see, for example, Barton 2010: 103–123, Bowater and Yoman 2013: 85–90). They must also undergo changes in their attitude to popular science as a genre. Those who write for lay audiences must embrace the literary features of popularization and not place all the emphasis on the information-carrying properties of their texts. Fictionality, which in popular science manifests through dramatization of communications and events involving scientists (in other words through presented voices), is a vital aspect that should not be ignored. Popular science has many features that it shares with literary fictions while remaining an information-dense genre. Therefore, it is vital for popular science authors to embrace the character-creating properties of presented discourse.

Similarly to the use of presented discourse, narrative structures employed in popular science are designed to showcase science and scientists as the forces of goodness. Popular science in many ways preserves and perpetuates the ideology of a professional research article. Harré's (1994: 86) observations about the nature of scientific writing show that popular science authors do not so much invent the personae of the idealized scientists as slightly exaggerate the norms of professional texts as Harré (1994: 86) interprets them. For instance, Harré (1994: 86) argues for strong elements of literary narrativity in professional research articles. He shows that the familiar structure of a research article (Introduction, Methods and Materials, Results, Discussion) can be reinterpreted to follow that of a literary narrative. Harré (1994: 86) calls the participant in a professional narrative a "hero." He goes on to say that this hero follows three predictable stages: first, he "presents a hypothesis" (Introduction), then he tests the hypothesis and obtains empirical support (Methods and Materials), and finally the hero presents the results of his experiments as proof for the initial hypothesis (Result and Discussion).

Not only is the narrative structure at work here, but it is used in such a way as to purge any negative results or problems with the experimental proof. Essentially, this is the narrative makeup proposed by Hoey's (2001) model. Harré (1994: 86) calls such presentation of science a "tale," "fiction," and a "bit of theater," meaning that in real-life situations, problems and negative outcomes do happen, and often. He also notes that any professional scientist would recognize this form of information organization as "a smiling face" of science, but would nevertheless follow these conventions in professional publications. Popular science takes the application of narrative structure one step further with pattern recycling and the solution preview options. These mechanisms ensure the positive outcome of any popular science narrative where they are used. They also train the reader to approach stories of scientific endeavors with an exclusively positive mind set.

Those who write popular science in book form should be aware of the applications of presented discourse and narrative structure specific to popularizations. It is important for authors not only to use these mechanisms correctly but also to be aware of their hidden potential.

Bibliography

Primary Texts

Bryson, B. (2003) *A Short History of Nearly Everything*. New York: Broadway Books.

Bynum, W. (2012) *A Little History of Science*. London: Yale University Press.

Carroll, S. (2012) *The Particle at the End of the Universe: How the Hunt for the Higgs Boson Leads Us to the Edge of a New World*. New York: Plume.

Coen, E. (2012) *Cells to Civilization: The Principles of Change that Shape Life*. Princeton: Princeton University Press.

du Sautoy, M. (2010) *The Number Mysteries: A Mathematical Odyssey through Everyday Life*. New York: Palgrave Macmillan.

Ferris, T. (1988) *Coming of Age in the Milky Way*. New York: William Morrow and Company.

Greene, B. (2011) *The Hidden Reality: Parallel Universes and the Deep Laws of the Cosmos*. New York: Alfred A. Knopf.

Kaku, M. (2011) *Physics of the Future: How Science Will Shape Human Destiny and Our Daily Lives by the Year 2100*. New York: Doubleday.

Kean, S. (2012) *The Violinist's Thumb and Other Lost Tales of Love, War, and Genius, as Written by Our Genetic Code*. New York: Little, Brown and Company.

Zimmer, C. (2011) *A Planet of Viruses*. Chicago: University of Chicago Press.

General Bibliography

Anderson, L.W. and D.R. Krathwohl (2001) *A Taxonomy for Learning, Teaching, and Assessing: A revision of Bloom's Taxonomy of Educational Objectives*. New York, NY: Longman.

Baerger, D.R. and D.P. McAdam (1999) "Life story coherence and its relation to psychological well-being" *Narrative Inquiry* 9/1: 69–96.

Banerjee, A. (2012) *We Modern People: Science Fiction and the Making of Russian Modernity*. Middletown, Connecticut: Wesleyan University Press.

Barberousse, A. and P. Ludwig (2009) "Models as fictions" in *Fictions in Science: Philosophical Essays on Modeling and Idealization*. Suarez, M. (ed.). London: Routledge: 56–73.

Barnes, J.L. (2015) "Fanfiction as imaginary play: what fan-written stories can tell us about the cognitive science of fiction" *Poetics* 48: 69–82. DOI 10.1016/j.poetic.2014.12.004

Baron, N. (2010) *Escape from the Ivory Tower: A Guide to Making Your Science Matter*. London: Island Press.

Barwich, A.-S. (2013) "Science and fiction: analysing the concept in science and its Limits" *Journal for General Philosophy of Science* 44: 357–373. DOI 10.1007/s10838-013-9228-2

Baynham, M. (1996) "Direct speech: what's it doing in non-narrative discourse?" *Journal of Pragmatics* 25: 61–81. DOI 10.1016/0378-2166(94)00074-3

Bell, A. (1984) "Language style as audience design" *Language in Society* 13/2: 145–204.

Bell, A. (1991) *The Language of News Media*. Oxford: Blackwell.

Bensaude-Vincent, B. (2001) "A genealogy of the increasing gap between science and the public" *Public Understanding of Science* 10: 99–113.

Berman, R.A. (1997) "Narrative theory and narrative development: The Labovian impact" *Journal of Narrative and Life History* 7/1–4: 235–244.

Blanchard, L., V. Bhaskar, and L. Briefer (2015) "The lost narrative: Ecosystem service narratives and the missing Wasatch watershed conservation story" *Ecosystem Services* 16: 105–111. DOI 10.1016/j.ecoser.2015.10.019

Bloom, B.S., M.D. Engelhart, E.J. Furst, W.H. Hill, and D.R. Krathwohl (1956) *Taxonomy of Educational Objectives: The Classification of Educational Goals*. New York: David McKay Company.

Bowater, L. and K. Yeoman (2013) *Science Communication: A Practical Guide for Scientists*. Chichester: Wiley-Blackwell.

Bucchi, M. (1998) *Science and the Media: Alternative Routes in Scientific Communication*. London: Routledge.

Calsamiglia, H. (2003) "Popularization discourse" *Discourse Studies* 5/2: 139–146.

Calsamiglia, H. and C.L. Ferrero (2003) "Role and position of scientific voices: reported speech in the media" *Discourse Studies* 5/2: 147–173.

Caracciolo, M. (2013) "Phenomenological metaphors in reader's engagement with characters: the case of Ian McEwans *Saturday*" *Language and Literature* 22/1: 60–76. DOI 10.1177/0963947012462948

Cartwright, N. (1983) *How the Laws of Physics Lie*. Oxford: Oxford University Press.

Ciapuscio, G. (2003) "Formulation and reformulation procedures in verbal interactions between experts and (semi-)laypersons" *Discourse Studies* 5/2: 207–233.

Cohn, D. (1990) "Signposts of fictionality: a narratological perspective" *Poetics Today* 11/4: 775–804.

Coppola, A. (2016) *The Theater of Experiment: Staging Natural Philosophy in Eighteenth-Century Britain*. Oxford: Oxford University Press.

Culler, J. (1982/2008) *On Deconstruction: Theory and Criticism after Structuralism*. Ithaca: Cornell University Press.

Dahl, T. (2015) "Contested science in the media: linguistic traces of news' writers framing activity" *Written Communication* 32/1: 39–65. DOI 10.1177/0741088314557623

Dawson, P. (2015) "Ten theses against fictionality" *Narrative* 23/1: 74–100. DOI 10.1353/nar.2015.0006 de Oliveira, J. and A. Pagano (2006) "The research article and the science popularization article: a probabilistic functional grammar perspective on direct discourse presentation" *Discourse Studies* 8/5: 627–646. DOI 10.1177/1461445606064833

Djerassi, C. (1994) *The Bourbaki Gambit*. New York: Penguin Books.

Eggins, S. and D. Slade (1997) *Analyzing casual conversation*. London: Cassell.

Fleischman, S. (1997) "The 'Labovian model' revisited with special consideration of literary narrative" *Journal of Narrative and Life History* 7/1–4: 159–168.

Fairclough, N. (1989/2013) *Language and Power*. London: Routledge.

Fine, A. (1993) "Fictionalism" *Midwest Studies in Philosophy* 18: 1–18.

Fiske, S., A. Cuddy, P. Glick, and J. Xu (2002) "A model of (often mixed) stereotype content: competence and warmth respectively follow from perceived status and competition" *Journal of Personal and Social Psychology* 82/6: 878–902. DOI 10.1037//0022-3514.82.6.878

Fludernik, M. (1993) *The Fictions of Language and the Languages of Fiction: The Linguistic Representation of Speech and Consciousness*. London: Routledge.

Fludernik, M. (1996) *Towards a 'Natural' Narratology*. New York: Routledge.

Fludernik, M. (2011) "The category of 'person' in fiction: *You* and *we* narrative-multiplicity and indeterminacy of reference" In *Current Trends in Narratology* Olson, G. (ed.) Berlin: de Gruyter: 101–141.

Fu, X. and K. Hyland (2014) "Interaction in two journalistic genres: a study of interactional metadiscourse" *English Text Construction* 7/1: 122–144. DOI 10.1075/etc.7.1.05fu

Gilbert, G. and M. Mulkay (1984) *Opening Pandora's Box: A Sociological Analysis of Scientists' Discourse*. Cambridge: Cambridge University Press.

Goffman, E. (1981) *Forms of Talk*. Philadelphia: University of Pennsylvania Press.

Goodman, A. (2006) *Intuition*. New York: The Dial Press.

Gordin, M.D. *Scientific Babel: How Science was Done Before and After Global English*. Chicago: University of Chicago Press.

Gülich, E. (2003) "Conversational techniques used in transferring knowledge between medical experts and non-experts" *Discourse Studies* 5/2: 235–263.

Halliday, M.A.K. and C.M.I.M. Matthiessen (2014) *Halliday's Introduction to Functional Grammar*. London: Routledge.

Harré, R. (1994) "Some narrative conventions of scientific discourse" in *Narrative in Culture: The Uses of Storytelling in Sciences, Philosophy, and Literature*. Nash, C. (ed.). London: Routledge: 81–101.

Harris, R. (2008) "The grammar of numbers" *Integrationist Notes and Papers* 19. Accessible: http://www.royharrisonline.com/grammar_of_numbers.html. Accessed: 16th May 2017.

Herman, D. (2009) *Basic Elements of Narrative*. Hoboken, NJ: Wiley-Blackwell.

Hermwille, L. (2016) "The role of narratives in socio-technical transitions—Fukushima and the energy regimes of Japan, Germany, and the United Kingdom" *Energy Research and Social Science* 11: 237–246. DOI 10.1016/j.erss.2015.11.001

Hoey, M. (1983) *On the Surface of Discourse*. London: George Allen and Unwin.

Hoey, M. (2001) *Textual Interaction: An Introduction to Written Discourse Analysis*. London: Routledge.

Hunston, S. (2000) "Evaluation and the planes of discourse: Status and value in persuasive texts" in *Evaluation in Text: Authorial Stance and the Construction of Discourse*. Hunston, S. and Thompson, G. (eds.). Oxford: Oxford University Press: 176–207.

Hunston, S. and G. Thompson (2000) *Evaluation in Text: Authorial Stance and the Construction of Discourse*. Oxford: Oxford University Press.

Hyland, K. (2001) "Bringing in the reader: addressee features in academic articles" *Written Communication* 18/4: 549–574.

Hyland, K. (2005) "Representing readers in writing: student and expert practices" *Linguistics and Education* 16: 363–377. DOI 10.1016/j.linged.2006.05.002

Hyland, K. (2009) "Corpus informed discourse analysis: the case of academic Engagement" in *Academic Writing: At the Interface of Corpus and Discourse*. Charles, M., D. Pecorari, and S. Hunston (eds.). London: Continuum: 110–128.

Hyland, K. (2010) "Constructing proximity: relating to readers in popular and professional science" *Journal of English for Academic Purposes* 9: 116–127. DOI 10.1016/j.jeap.2010.02.003

Iser, W. (1972) *The Implied Reader: Patterns of Communication in Prose Fiction from Bunyan to Beckett*. Baltimore: John Hopkins University Press.

Kacandes, I. (2001) *Talk Fiction: Literature and the Talk Explosion*. Lincoln: University of Nebraska Press.

Keene, M. (2014) "Familiar science in nineteenth-century Britain" *History of Science*, 52/1: 53–71.

Koteyko, N., B. Nerlich, P. Crawford, and N. Wright (2008) "'Not rocket science' or 'no silver bullet?' Media and government discourses about MRSA and cleanliness" *Applied Linguistics* 29/2: 223–243. DOI 10.1093/applin/amn006

Kranich, S. (2016) *Contrastive Pragmatics and Translation. Evaluation, Epistemic Modality and Communicative Style in English and German*. Amsterdam: John Benjamins.

Kranich, Svenja, Juliane House and Viktor Becher (2012) "Changing conventions in English-German translations of popular scientific texts" In *Multilingual Individuals and Multilingual Societies*. Braunmüller, K. and C. Gabriel (eds.) Amsterdam: John Benjamins: 315–334.

Krathwohl, D.R. (2002) "A revision of Bloom's taxonomy: an overview" *Theory into Practice* 41/4: 212–264.

Kylinich, M.A. & Pokalyhina, M.V. (2013) "About dialogicity of paratext elements 'introduction/conclusion' in English language popular science" *News of the Samara Scientific Center of the Russian Academy of Sciences* 2/3: 776–770

Labov, W. (1972) *Language in the Inner City: Studies in the Black English Vernacular*. Philadelphia: University of Pennsylvania Press.

Labov, W. (1997) "Some further steps in narrative analysis" *Journal of Narrative and Life History* 7/1–4: 395–415.

Labov, W. and J. Waletsky (1967) "Narrative analysis: oral versions of personal experience" in *Essays on the Verbal and Visual Arts: Proceedings of the 1966 Annual Spring Meeting of the American Ethnological Society*. Helm, J. (ed.). Seattle: University of Washington Press: 12–44.

Lamarque, P. (2000) "Fiction and reality" in *Philosophy and Fiction: Essays in Literary Aesthetics*. Lamarque, P. (ed.). Christchurch, New Zealand: Cybereditions Corporation: 71–92.

Laslo, E., A. Baram-Tsabari, and B. Lewenstein (2011) "A growth medium for the message: online science journalism affordances for exploring public discourse of science and ethics" *Journalism* 12/7: 847–870. DOI 10.1177/1464884911412709

Leech, G. and M. Short (1981/2007) *Style in Fiction: A Linguistic Introduction to English Fictional Prose*. London: Pearson.

Lerer, S. (2015) *Inventing English: A Portable History of the Language*. New York: Columbia University Press.

Lewin, B. and H. Perpignan (2012) "Recruiting the reader in literary criticism" *Text and Talk* 32/6: 751–772. DOI 10.1515/text-2012-0035

Lightman, B. (2000) "Marketing knowledge for the general reader: Victorian popularizers of science" *Endeavour* 24/3: 100–106.

Livnat, Z. (2012) *Dialogue, Science and Academic Writing*. Amsterdam: John Benjamins Publishing Company.

Luzón, M. (2013) "Public communication of science in blogs: recontextualizing scientific discourse for a diversified audience" *Written Communication* 30/4: 428–457. DOI 10.1177/0741088313493610

Mair, C. (2006) *Twentieth-century English: History, Variation, and Standardization*. Cambridge: Cambridge University Press.

Martin, J.R. and P.R.R. White (2005) *The language of Evaluation: Appraisal in English*. New York: Palgrave Macmillan.

Meredith, D. (2010) *Explaining Research: How to Reach Key Audiences to Advance Your Work*. Oxford: Oxford University Press.

Mildorf, J. (2008) "Thought presentation and constructed dialogue in oral stories: limits and possibilities of a cross-disciplinary narratology" *Partial Answers: Journal of Literature and the History of Ideas* 6/2: 279–300. DOI 10.1353/pan.0.0025

Mildorf, J. (2016) "Reconsidering second-person narration and involvement" *Language and Literature* 25/2: 145–158. DOI 10.1177/0963947016638985

Moirand, S. (2000) "Variations discursives dans deux situations productrices d'un discourse sur la science de la presse ordinaire" *Carnets du Cediscor* 6: 45–62.

Moirand, S. (2003) "Communicative and cognitive dimensions of discourse on science in the French mass media" *Discourse Studies* 5/2: 175–206.

Molek-Kozakowska, K. "Framing disease, aging, and death in Popular Science Journalism" *Brno Studies in English* 42/1: 49–69. DOI: 10.5817/BSE2016-1-3

Murphy, T. (2005) "The uncertainties of conversational exchange: dialogue monitoring as a function of the narrative voice" *Style* 39/4: 396–411.

Myers, G. (1990). *Writing Biology: Texts in the Social Construction of Scientific Knowledge.* Madison, Wisconsin: The University of Wisconsin Press.

Myers, G. (1992) "Fictions for facts: the form and authority of the scientific dialogue" *History of Science* 30: 221–247.

Myers, G. (1997) "Fictionality, demonstration, and a forum for popular science: Jane Marcet's *Conversations on Chemistry*" in *Natural Eloquence: Women Reinscribe Science.* Gates, B.T. and A.B. Shteir (eds.) Madison: University of Wisconsin Press: 43–60.

Myers, G. (1999) "Functions of reported speech in group discussions" *Applied Linguistics* 20/3: 376–401.

Myers, G. (2003) "Discourse studies of scientific popularisation: questioning the Boundaries" *Discourse Studies* 5/2: 265–279.

Myers, G. (2010) *Discourse of blogs and wikis.* London: Continuum International Publishing.

Olson, R. (2015) *Houston, We Have a Narrative: Why Science Needs Story.* Chicago: The University of Chicago Press.

Oparina, O.I. (2015) "Science, worldview, and language in the XIX century (some developments and basic tendencies)" *Philology and Person* 2: 63–74.

Oxford English Dictionary Online (2014) Ed. John Simpson. Oxford: Oxford University Press. Accessible: http://www.oed.com/ Accessed: 7th May 2017.

Parkinson, J. and R. Adendorff (2004) "The use of popular science articles in teaching scientific literacy" *English for Specific Purposes* 23: 379–396. DOI 10.1016/j.esp.2003.11.005

Peterson, C. and A. McCabe (1983) *Developmental Psycholinguistics: Three Ways of Looking at a Child's Narrative.* London: Plenum Press.

Perrault, S. (2008) "The cons of the pros: how popular science writers view their readers." Paper presented at 62nd Annual RMMLA Convention—Rocky Mountain Modern Language Association. Reno, Nevada, USA, 9–11 October.

Pilkington, A. (2017) *Science Fiction and Futurism: Their Terms and Ideas.* Jefferson, North Carolina: McFarland.

Pilkington, O. (2016) "Popular science as a means of emotional engagement with the scientific community" *International Journal of Science Culture and Sport* 4/1: 118–125. DOI 10.14486/IntJSCS466

Pilkington, O. (2017a) "Popular science versus lab lit: Differently depicting scientific apparatus" *Science as Culture*, 14 Nov: 1–22. DOI 10.1080/09505431.2016.1255722

Pilkington, O. (2017b) "Structural complexity of popular science narratives of discovery as an indicator of reader-awareness: A Labov-inspired approach," *Linguistic and Philosophical Investigations* 16: 7–28. DOI 10.22381/LPI1620171

Pilkington, O. (2018) "Presented discourse in popular science narratives of discovery: Communicative side of thought presentation" *Linguistic and Philosophical Investigations* 17: 7–28. DOI 10.22381/LPI1720181

Pilkington, O. and A. Pilkington eds. (forthcoming) *Lab Lit: Exploring Literary Fictions about Science*. Lanham, Maryland: Lexington Books.

Reitsma, F. (2010) "Geoscience explanations: Identifying what is needed for generating scientific narratives from data models." *Environmental Modelling & Software* 25: 93–99. DOI 10.1016/j.envsoft.2009.07.011

Rimmon-Kenan, S. (2002) *Narrative Fiction*. London: Routledge.

Roberts, K. (2013) "Foreword" in *Science Communication: A Practical Guide for Scientists*. Bowater, L. and K. Yeoman (eds.). Chichester: Wiley-Blackwell.

Rohn, J. (2005) "What is lab lit (the genre)?" Accessible: http://www.lablit.com/article/3. Accessed: 5th March 2017.

Rohn, J. (2010a) "More lab in the library" *Nature* 465/3: 552.

Rohn, J. (2010b) *The Honest Look*. Cold Spring Harbor, New York: Cold Spring Harbor Laboratory Press.

Rouse, J. (2009) "Laboratory fictions" in *Fictions in Science: Philosophical Essays on Modeling and Idealization*. Suarez, M. (ed.). London: Routledge: 37–55.

Sackler, M. (ed.) (2014) *Science of Science Communication II: Summary of a Colloquium*. Washington, D.C.: The National Academies Press.

Schekman, R. (2013) "How journals like *Nature*, *Cell* and *Science* are damaging science" *The Guardian*, 9th December. 2013. Accessible: https://www.theguardian.com/commentisfree/2013/dec/09/how-journals-nature-science-cell-damage-science. Accessed: 2nd June 2017.

Semino, E. (2004) "Representing characters' speech and thought it narrative: a study of England, England by Julian Barnes" *Style* 38/4: 428–451.

Semino, E. and M. Short (2004) *Corpus Stylistics: Speech, Writing and Thought Presentation in a Corpus of English Writing*. London: Routledge.

Short, M. (1988) "Speech presentation, the novel and the press" in (1988) *The Taming of the Text: Explorations in Language, Literature, and Culture*. Van Peer (ed.). Cambridge: Cambridge University Press: 63–81.

Short, M. (2007) "Thought presentation twenty-five years on" *Style* 41/2: 225–241.

Short, M. (2012) "Discourse presentation and speech (and writing but not thought) summary" *Language and Literature* 21/1: 18–32. DOI 10.1177/0963947011432049

Short, M., E. Semino, and M. Wynne (2002) "Revisiting the notion of faithfulness in discourse presentation using a corpus approach" *Language and Literature* 11/4: 325–355.

Skov Nielsen, H., J. Phelan, and R. Walsh (2015 a) "Ten theses about fictionality" *Narrative* 23/1: 61–73. DOI 10.1353/nar.2015.0005

Skov Nielsen, H., J. Phelan, and R. Walsh (2015 b) "Fictionality as rhetoric: a response to Paul Dawson" *Narrative* 23/1: 101–111. DOI 10.1353/nar.2015.0000

Smirnova, A.V. (2009) "Reported speech as an element of argumentative newspaper discourse" *Discourse & Communication* 3/1: 79–103. DOI: 10.1177/1750481308098765

Stanley, M. (2011) "How scientists stopped talking about science" *Studies in History and Philosophy of Science* 42: 235–239. DOI 10.1016/j.shpsa.2010.11.015

Suarez, M. (2009) "Fictions in scientific practice" *Fictions in Science: Philosophical Essays on Modeling and Idealization*. Suarez, M. (ed.). London: Routledge: 3–15.

Supper, A. (2014) "Sublime frequencies: the construction of sublime listening experiences in the sonification of scientific data" *Social Studies of Science* 44/1: 34–58. DOI 10.1177/0306312713496875

Tannen, D. (2007) *Talking Voices: Repetition, Dialogue, and Imagery in Conversational Discourse*. Cambridge: Cambridge University Press.

Talbot, M. (1995) "A synthetic sisterhood: false friends in a teenage magazine" in *Gender Articulated: Language and the Socially Constructed Self*. Hall, K. and M. Bucholtz (eds.). New York: Routledge: 143–168.

Thompson, G. (1996) "Voices in the text: discourse perspective on language reports" *Applied Linguistics* 17/4: 501–530.

Thompson, G. (2001) "Interaction in academic writing: learning to argue with the Reader" *Applied Linguistics* 22/1: 58–78.

Thompson, G. (2012) "Intersubjectivity in newspaper editorials: constructing the reader-in-the-text" *English Text Construction* 5/1: 77–100.

Thompson, G. and P. Thetela (1995) "The sound of one hand clapping: the management of interaction in written discourse" *Text* 15/1: 103–127. DOI 10.1075/etc.5.1.05tho

Toolan, M. (2001) *Narrative: A Critical Linguistic Introduction*. London: Routledge.

Toolan, M. (2011). "The texture of emotionally-immersive passages in short stories: steps towards a tentative local grammar" in *Proceedings of the 3rd International Stylistics Conference*. Accessible: http://artsweb.bham.ac.uk/MToolan/UPLOADS/MICHAEL%20TOOLAN%20Ningbo%20March%202011%20-%20Local%20Grammar.pdf Accessed: 16th September 2014.

Toon, A. (2012) *Models and Make-Believe: Imagination, Fiction and Scientific Representation*. New York: Palgrave McMillan.

Topham, J. (2000) "Scientific publishing and the reading of science in nineteenth-century Britain: a historiographical survey and guide to sources" *Studies in History and Philosophy of Science* 31/4: 559–612.

Turney, J. (2004a) "The abstract sublime: life as information waiting to be rewritten" *Science as Culture* 13/1: 89–103. DOI 10.1080/0950543042000193799

Turney, J. (2004b) "Accounting for explanation in popular science texts—an analysis of popularized accounts of superstring theory" *Public Understanding of Science* 13:331–346. DOI 10.1177/0963662504044909

Turney, J. (2007) "Boom and bust in popular science" *Journal of Science Communication* 6/1: 1–3.

Urbanova, Z. (2012) "Direct and free direct forms of representation in the discourse of newspaper reports: less frequent phenomena" *Brno Studies in English* 38/1: 39–54. DOI 10.5817/BSE2012-1-3

Vaihinger, H. (1924/2001) *The Philosophy of 'As If'* C.K. Ogden (trans.) London: Routledge.

Votrina, E.N. (2012) *The Function of Dialogicity in Popular Science of the 20th Century*. Volgograd, Russian Federation, PhD Thesis.

Waugh, L. (1995) "Reported speech in journalistic discourse: the relation of function and text" *Text* 15/1: 129–173.

Yang, B. (2003) "Toward a holistic theory of knowledge and adult learning" *Human Resource Development Review* 2/2: 106–129. DOI 10.1177/1534484303002002002

Index of Subjects and Names

www.ingramcontent.com/pod-product-compliance
Ingram Content Group UK Ltd.
Pitfield, Milton Keynes, MK11 3LW, UK
UKHW021421091026
15624UKWH00010B/330

9 789004 365964